# FORENSIC ANALYSIS

# WEIGHING BULLET LEAD EVIDENCE



Committee on Scientific Assessment of Bullet Lead Elemental Composition Comparison

Board on Chemical Sciences and Technology

Division of Earth and Life Studies

NATIONAL RESEARCH COUNCIL
*OF THE NATIONAL ACADEMIES*

THE NATIONAL ACADEMIES PRESS
Washington, D.C.
**www.nap.edu**

**THE NATIONAL ACADEMIES PRESS 500 Fifth Street, NW Washington, DC 20001**

NOTICE: The project that is the subject of this report was approved by the Governing Board of the National Research Council, whose members are drawn from the councils of the National Academy of Sciences, the National Academy of Engineering, and the Institute of Medicine. The members of the committee responsible for the report were chosen for their special competences and with regard for appropriate balance.

Support for this study was provided by the Federal Bureau of Investigation under Contract No. S2N0216700.

Any opinions, findings, conclusions, or recommendations expressed in this publication are those of the author(s) and do not necessarily reflect the views of the organizations or agencies that provided support for the project.

International Standard Book Number 0-309-09079-2 (Book)
International Standard Book Number 0-309-52756-2 (PDF)

Library of Congress Catalog Card Number 2004101584

A limited number of copies of the report are available from the Board on Chemical Sciences and Technology, 500 Fifth Street, NW, Washington, DC 20001.

Additional copies of this report are available from the National Academies Press, 500 Fifth Street, N.W., Box 285, Washington, DC 20055; (800) 624-6242 or (202) 334-3313 (in the Washington metropolitan area); http://www.nap.edu



Printed in the United States of America

# THE NATIONAL ACADEMIES

***Advisers to the Nation on Science, Engineering, and Medicine***

The **National Academy of Sciences** is a private, nonprofit, self-perpetuating society of distinguished scholars engaged in scientific and engineering research, dedicated to the furtherance of science and technology and to their use for the general welfare. Upon the authority of the charter granted to it by the Congress in 1863, the Academy has a mandate that requires it to advise the federal government on scientific and technical matters. Dr. Bruce M. Alberts is president of the National Academy of Sciences.

The **National Academy of Engineering** was established in 1964, under the charter of the National Academy of Sciences, as a parallel organization of outstanding engineers. It is autonomous in its administration and in the selection of its members, sharing with the National Academy of Sciences the responsibility for advising the federal government. The National Academy of Engineering also sponsors engineering programs aimed at meeting national needs, encourages education and research, and recognizes the superior achievements of engineers. Dr. Wm. A. Wulf is president of the National Academy of Engineering.

The **Institute of Medicine** was established in 1970 by the National Academy of Sciences to secure the services of eminent members of appropriate professions in the examination of policy matters pertaining to the health of the public. The Institute acts under the responsibility given to the National Academy of Sciences by its congressional charter to be an adviser to the federal government and, upon its own initiative, to identify issues of medical care, research, and education. Dr. Harvey V. Fineberg is president of the Institute of Medicine.

The **National Research Council** was organized by the National Academy of Sciences in 1916 to associate the broad community of science and technology with the Academy's purposes of furthering knowledge and advising the federal government. Functioning in accordance with general policies determined by the Academy, the Council has become the principal operating agency of both the National Academy of Sciences and the National Academy of Engineering in providing services to the government, the public, and the scientific and engineering communities. The Council is administered jointly by both Academies and the Institute of Medicine. Dr. Bruce M. Alberts and Dr. Wm. A. Wulf are chair and vice chair, respectively, of the National Research Council.

**www.national-academies.org**

## COMMITTEE ON SCIENTIFIC ASSESSMENT OF BULLET LEAD ELEMENTAL COMPOSITION COMPARISON

KENNETH O. MACFADDEN, Independent Consultant, Chair
A. WELFORD CASTLEMAN, JR., The Pennsylvania State University
PETER R. DE FOREST, John Jay College of Criminal Justice
M. BONNER DENTON, University of Arizona
CHARLES A. EVANS, JR., Consultant
MICHAEL O. FINKELSTEIN, Attorney
PAUL C. GIANNELLI, Case Western Reserve University
ROBERT R. GREENBERG, National Institute of Standards and Technology
JAMES A. HOLCOMBE, University of Texas
KAREN KAFADAR, University of Colorado at Denver
CHARLES J. MCMAHON, JR., University of Pennsylvania
STEVEN R. PRESCOTT, Hercules, Inc.
CLIFFORD SPIEGELMAN, Texas A&M University
RAYMOND S. VOORHEES, United States Postal Inspection Service

**Staff**

JENNIFER J. JACKIW, Program Officer
SYBIL A. PAIGE, Administrative Associate
DAVID C. RASMUSSEN, Program Assistant
DOROTHY ZOLANDZ, Director, Board on Chemical Sciences and Technology
MICHAEL COHEN, Senior Program Officer, Committee on National Statistics

## BOARD ON CHEMICAL SCIENCES AND TECHNOLOGY

# Acknowledgment of Reviewers

This report has been reviewed in draft form by individuals chosen for their diverse perspectives and technical expertise, in accordance with procedures approved by the National Research Council's Report Review Committee. The purpose of this independent review is to provide candid and critical comments that will assist the institution in making the published report as sound as possible and to ensure that the report meets institutional standards for objectivity, evidence, and responsiveness to the study charge. The review comments and draft manuscript remain confidential to protect the integrity of the deliberative process. We wish to thank the following individuals for their review of this report:

Ramon M. Barnes, University of Massachusetts, Amherst
Charles K. Bayne, Oak Ridge National Laboratory
Margaret A. Berger, Brooklyn Law School
Steven D. Brown, University of Delaware
Keith Eberhardt, Kraft Foods
Kenneth D. Green, Sporting Arms and Ammunition Manufacturers' Association
Kenneth N. Han, South Dakota School of Mines and Technology
Brent Hiskey, University of Arizona
Alan Karr, National Institute of Statistical Sciences
Kenneth Kees, ATK
John A. Koropchak, Southern Illinois University, Carbondale
Steven R. Moore, ATK
R. David Prengaman, RSR Technologies, Inc.
Walter F. Rowe, The George Washington University
Alan Serven, Remington Arms Co., Inc.
William A. Tobin, Forensic Metallurgical Consultant

Gregory C. Turk, National Institute of Standards and Technology
Diarmuid White, Law Firm of White & White

Although the reviewers listed above have provided many constructive comments and suggestions, they were not asked to endorse the conclusions or recommendations nor did they see the final draft of the report before its release. The review of this report was overseen by **Hyla S. Napadensky,** Napadensky Energetics, Inc. (retired), and **Royce W. Murray,** University of North Carolina. Appointed by the National Research Council, he was responsible for making certain that an independent examination of this report was carried out in accordance with institutional procedures and that all review comments were carefully considered. Responsibility for the final content of this report rests entirely with the authoring committee and the institution.

# Preface

This study was initiated by discussions between the Federal Bureau of Investigation (FBI) and National Research Council staff. Because compositional analysis of bullet lead (CABL) has recently come under greater scrutiny, the FBI desired an impartial scientific assessment of the soundness of the scientific principles underlying CABL to determine the optimum manner for conducting the examination and to establish scientifically valid conclusions that can be reached using the examination. After the development of a feasible statement of task, a committee that had the expertise required by the statement of task was assembled. The nominees underwent the National Research Council's rigorous nomination process before approval was given, to identify any bias or conflict of interest prior to the start of the project.

The committee met four times—once a month—beginning in February 2003 (the meeting agendas are found in Appendix C). This demanding schedule was met by the committee members with positive attitudes, and the effort put forth to review journal articles and trial transcripts, run statistical tests, and produce this report was tremendous.

Sincere thanks are offered to many others who provided the committee with information on the intricacies of the issues surrounding the study. Space does not permit naming of all who contributed, but some individuals who were particularly helpful are mentioned here. Representatives of the FBI, especially Robert Koons, attended the open session at every meeting to answer the committee's many questions. Diana Grant, also of the FBI, was kind enough to take the time to demonstrate the process of comparative bullet lead analysis from start to finish as part of a laboratory tour. All of the speakers who gave presentations at the committee meetings are greatly appreciated for taking the time to assist the committee with this matter of national importance. Special thanks go to Kenneth Green of the Sporting Arms and Ammunition Manufacturers' Association for his repeated support throughout the study. Frederick Whitehurst and William Tobin have willingly shared their extensive collections of legal documents with the cmmittee. Troy Roseberry of PMC-Eldorado Cartridge Company is deserving of thanks for guiding the committee through the ammunition production process at the Boulder City, Nevada, facility. Finally, John Bailar, Scholar-in-Residence at the National Academies, was invaluable for his assistance and insights into the statistical aspects of this report.

I thank everyone who helped further the successful completion of this study.

Kenneth O. MacFadden, *Chair*
Committee on Scientific Assessment of
Bullet Lead Elemental Composition Comparison

# Contents

EXECUTIVE SUMMARY 1

1 INTRODUCTION 8

2 COMPOSITIONAL ANALYSIS 12

3 STATISTICAL ANALYSIS OF BULLET LEAD DATA 26

4 INTERPRETATION 71
Significance of the Bullet Manufacturing Process, 71
Compositional Analysis of Bullet Lead as Evidence in the Legal System, 85

5 MAJOR FINDINGS AND RECOMMENDATIONS 109

APPENDIXES

A Statement of Task 117
B Committee Membership 118
C Committee Meeting Agendas 124
D Glossary 130
E Basic Principles of Statistics 133

| | | |
|---|---|---|
| F | Simulating False Match Probabilities Based on Normal Theory | 142 |
| G | Data Analysis of Table 1, Randich et al. | 151 |
| H | Principal Components Analysis: How Many Elements Should Be Measured? | 157 |
| I | Birthday Problem Analogy, | 163 |
| J | Understanding the Significance of the Results of Trace Elemental Analysis of Bullet Lead | 165 |
| K | Statistical Analysis of Bullet Lead Data by Karen Kafadar and Clifford Spiegelman | 169 |

# Executive Summary

When a crime involves gunfire, examination of physical evidence derived from ammunition often yields key pieces of evidence used in the investigation of that crime. Firearms examination focuses on characteristic marks left on fired bullets and expended cartridge cases by the weapon from which the cartridge is discharged. With bullets, this involves matching the striations on a bullet caused by its passage through the barrel of a gun with marks on test bullets fired through the barrel of a gun found in the possession of a suspect. However, frequently, no gun is recovered, or a bullet fragment is too small or mangled to observe adequate striations. In such instances, a different approach must be explored to evaluate the possibility of a link between the crime scene bullet(s)[1] and the suspect.

One such approach is compositional analysis of bullet lead (CABL), which has been used by the law-enforcement community to provide circumstantial evidence for criminal investigation and prosecution since the 1960s. Crime scene investigators and autopsy pathologists collect bullet fragments (and sometimes a bullet in its entirety) from a crime scene or the body of a victim in order to compare them with unused cartridges in the possession of a suspect (suspect's bullets) that investigators may have collected.

The FBI examiner takes three samples from each bullet or bullet fragment and analyzes them by a process known as inductively coupled plasma-optical emission spectroscopy (ICP-OES). This process is used to determine the concentrations of seven selected elements—arsenic (As), antimony (Sb), tin (Sn), copper (Cu), bismuth (Bi), silver (Ag), and cadmium (Cd)—in the bullet lead alloy of both the

[1] The term *crime scene bullet* includes bullet fragments and shot from shotguns. This evidence may be recovered at a crime scene or from a victim at a hospital or during an autopsy.

crime-scene and the suspect's bullets. The FBI examiner applies statistical tests to compare the elements in each crime-scene fragment with the elements in each of the suspect's bullets. If any of the fragments and suspect's bullets are determined statistically to be analytically indistinguishable for each of the elemental concentration means, the examiner's expert court testimony currently will indicate that the fragments and bullets probably came from the same "source."

The Federal Bureau of Investigation (FBI) asked the National Research Council to conduct an impartial scientific assessment of the soundness of the principles underlying CABL, the optimal manner for conducting an examination with CABL, and the scientifically valid conclusions that can be reached with CABL. In particular, the FBI asked the National Research Council to address the following three subjects and specific questions:

- *Analytical method.* Is the method analytically sound? What are the relative merits of the methods currently available? Is the selection of elements used as comparison parameters appropriate? Can additional useful information be gained by measurement of isotopic compositions?
- *Statistics for comparison.* Are the statistical tests used to compare two samples appropriate? Can known variations in compositions introduced in manufacturing processes be used to model specimen groupings and provide improved comparison criteria?
- *Interpretation issues.* What are the appropriate statements that can be made to assist the requester in interpreting the results of compositional bullet lead comparison, for both indistinguishable and distinguishable compositions? Can significance statements be modified to include effects of such factors as the analytical technique, manufacturing process, comparison criteria, specimen history, and legal requirements?

The committee's assessment of these questions and its overarching recommendations are summarized below. Its complete recommendations are found in the body of the report and collected in Chapter 5. The full report provides clear comments on the validity of the chemical and statistical analyses utilized in CABL, and on what can and cannot validly be stated in court regarding CABL evidence. It is up to prosecutors and judges to use the conclusions of this report to decide whether CABL evidence has enough value to be introduced in any specific case.

## ANALYTICAL METHODOLOGY

The current analytical instrumentation used by the FBI is appropriate and is the best available technology with respect to both precision and accuracy for the elements analyzed in a lead matrix. No other technique for this application provides as good or better quantitative, multi-element capability; wide linear dynamic range; limited interferences; and low (parts per billion) detection and quantitative limits. Furthermore, the elements selected by the FBI for analysis

(As, Sb, Sn, Cu, Bi, Ag, and Cd) are appropriate in the sense that they are quantifiable through the use of ICP-OES. Measurements of Sb, Sn, Cd, As, and Cu provide the best discrimination between bullets, and although measurements of Bi and Ag have less probative value, their measurement offers no disadvantage relative to the time and effort needed for analysis by ICP-OES. **Recommendation: The FBI should continue to measure the seven elements As, Sb, Sn, Cu, Bi, Ag, and Cd through ICP-OES as stated in the current analytical protocol. Also, the FBI should evaluate the potential gain from the use of high-performance ICP-OES because improvement in analytical precision may provide better discrimination.**

The committee also considered the use of approaches other than CABL to improve the ability to compare crime-scene evidence with a suspect's bullets. For example, it has been reported that lead isotope determination can provide the high-precision analysis necessary to differentiate and identify bullet samples made from ores from different mines. At this time the method in its most practical form has not been shown to be particularly effective for differentiating among United States-based sources of lead. However the method may prove useful in conjunction with the ICP-OES method should the amount of foreign ammunition in use in the United States increase.

Although the current analytical technique is sound, the FBI Laboratory's practices in quality assurance must be improved significantly to ensure the validity of its results. Chapter 2 includes detailed recommendations for how the FBI's analytical practices should be improved. For example, the laboratory's analytical protocol should be revised to contain all details of the procedure and to provide a better basis for the statistics of bullet comparison. The laboratory also needs to develop a more comprehensive formal and documented proficiency test of each examiner and carry out studies to quantify measurement repeatability and reproducibility. After they have been revised based on the recommendations in Chapter 2, the details of the FBI's CABL procedure and the research and data that supports it should be published in a peer-reviewed journal or at a minimum its analytical protocol should be made available through some other public venue. The revised procedures also must be used consistently within the FBI Laboratory. **Recommendation: The FBI's documented analytical protocol should be applied to *all* samples and should be followed by *all* examiners for *every* case.**

## STATISTICS FOR COMPARISON

The FBI's documented statistical protocol for matching CABL evidence[2] describes a statistical procedure known as "chaining." The chaining process

[2] C.A. Peters, "Comparative Elemental Analysis of Firearms Projectile Lead By ICP-OES," FBI Laboratory Chemistry Unit. Issue date: Oct. 11, 2002. *Unpublished* (2002).

compares each evidence bullet (both from the crime scene and from the suspect, and which cannot be eliminated based on physical comparison) to the next sequentially to identify compositional groups in which all bullets and fragments are analytically indistinguishable within 2 standard deviations of each element's average concentration. The standard deviation (SD) of each elemental concentration is determined on the basis of the variation found among all bullets and fragments analyzed for the particular case under investigation. If all seven of the concentration intervals (from mean – 2SD to mean + 2SD) of any of the crime-scene fragments fall within one of the compositional groups formed by the suspect's bullets, the fragments and matching suspect's bullets are stated to be "analytically indistinguishable."

In the committee's assessment, chaining may lead to artificially large compositional groups of analytically indistinguishable bullets, thus causing a crime-scene fragment and a suspect's bullet to fall within the same analytically indistinguishable compositional group when this would not be true if other statistical methods were used. In addition, because of the small amount of data in any one study, the standard deviation from the evidence in the case will most likely be larger, less reliable, and more variable than the standard deviation of the analytical method when calculated over many studies (with pooled data).

Although the chaining method is the FBI's documented statistical protocol, discussions with FBI staff led the committee to believe that the FBI is no longer using it. Instead, the unwritten protocol compares each of the crime-scene fragments with each individual suspect's bullet (not with a compositional group). This method, 2-standard deviation overlap, deems bullets to be analytically indistinguishable if the intervals (from mean – 2SD to mean + 2SD) for the seven elemental concentrations for a crime-scene bullet and a suspect's bullet overlap. The FBI claims based on analysis of historical data that this current procedure for bullet comparison will result in a false match probability (FPP) of 1 in 2,500. This report provides better methods for estimating false match and false non-match probabilities due to measurement error.

The full report examines the FBI's current statistical protocol and provides detailed recommendations about how it should be revised in order to provide a sound basis for determining whether crime-scene evidence and suspects' bullets are analytically indistinguishable. For example, within-bullet measurement standard deviations should be estimated using a pooled standard deviation over many bullets that have been analyzed with the same ICP-OES technique. In addition, a detailed statistical investigation of the FBI's historical data set containing 71,000 bullets should be conducted to confirm the validity of the revised statistical protocol and the accuracy of the values used to assess the measurement uncertainty in each element. The revised procedures also must be used consistently within the FBI Laboratory. **Recommendation: The committee recommends that the FBI use either the $T^2$ test statistic or the successive**

**t-test statistics procedure described in this report in place of the 2-SD overlap, range overlap, and chaining procedures. Recommendation: The FBI's statistical protocol should be properly documented and followed by *all* examiners in *every* case.**

## SIGNIFICANCE OF THE MANUFACTURING PROCESS IN THE INTERPRETATION OF EVIDENCE

The committee reviewed the lead bullet manufacturing process to determine whether known variations in lead compositions introduced in the manufacturing process can be used to improve CABL comparison data. In the United States, lead recycled primarily from car batteries is melted and refined at a secondary lead smelter to produce an intermediate lead ingot or billet. The ingot or billet is purchased by a bullet manufacturer and extruded into a large wire roll, which is cut to produce lead slugs whose length and diameter depend on the caliber of ammunition. Slugs are pressed into the form of a bullet and are stored in bins according to caliber. The slugs are sometimes molded into a thimble-shaped copper alloy cup to form a jacketed bullet and then loaded into a cartridge. Cartridges are boxed immediately by some manufacturers. Other manufacturers may store the cartridges in bins by caliber until a customer order must be filled, at which time boxes are filled with cartridges, stamped with a lot number, and collected in cases or pallets for shipment.

In practice, the detailed process followed by each manufacturer varies, and the process can vary even within a single manufacturer to meet demand. For example, many bullet manufacturers add scrap lead from the bullet production to the melt at random times, sporadically changing the composition of the original melt. Likewise, the binning of bullets and cartridges may introduce more mixing of bullets from different melts. In fact, the FBI's own research has shown that a single box of ammunition can contain bullets from as many as 14 distinct compositional groups. **Finding: Variations among and within lead bullet manufacturers make any modeling of the general manufacturing process unreliable and potentially misleading in CABL comparisons.**

The committee also reviewed testimony from the FBI regarding the identification of the "source" of crime-scene fragments and suspects' bullets. Because there are several poorly characterized processes in the production of bullet lead and ammunition, as well as ammunition distribution, it is very difficult to define a "source" and interpret it for legal purposes. It is evident to the committee that in the bullet manufacturing process there exists a volume of material that is compositionally indistinguishable, referred to by the committee as a "compositionally indistinguishable volume of lead" or CIVL. That volume could be the melt, sows, or billets, which vary greatly in size, or some subpart of these. One CIVL yields a number of bullets that are analytically indistinguishable. Those bullets may be packed in boxes with bullets from other similar (but distinguishable) volumes or in boxes with bullets from the same compositionally indistinguishable volume of lead.

The committee attempted to obtain information on the distribution of ammunition and bullets in the United States. Such distribution information would assist with determining the probability of finding a large number of analytically indistinguishable bullets in one geographic region. Thus, the probability that a crime scene bullet which matches a suspect's bullet actually came from the suspect might be vastly different in an isolated small town vs a major metropolitan area. But, distribution information on bullets and on loaded ammunition either does not exist or is considered proprietary, and the committee was unable to assess regional distribution patterns. For these reasons, unlike the situation with some forms of evidence such as DNA typing of bloodstains, it is not possible to obtain accurate and easily understood probability estimates that are directly applicable.

## Legal Interpretations

In legal proceedings, the interpretation of CABL results depends on the quality of the chemical analysis of the evidence bullets and bullet fragments, the statistical comparison of those bullets, and determination of the significance of the comparison. The committee found the analytical technique used is suitable and reliable for use in court, as long as FBI examiners apply it uniformly as recommended. The recommended changes in the statistical procedures would provide a sound basis for whether crime-scene evidence and a suspect's bullets "match," that is, whether they are analytically indistinguishable. However for legal proceedings, the probative value of these findings and how that probative value is conveyed to a jury remains a critical issue.

Despite the variations in manufacturing processes that make it difficult to determine whether bullets come from the same compositionally indistinguishable volume of lead (CIVL), CABL analysis can have value in some court cases. **Finding: The committee found that CABL is sufficiently reliable to support testimony that bullets from the same CIVL are more likely to be analytically indistinguishable than bullets from different CIVLs. An examiner may also testify that having CABL evidence that two bullets are analytically indistinguishable increases the probability that two bullets came from the same CIVL, versus no evidence of match status. Recommendation: Interpretation and testimony of examiners should be limited as described above, and assessed regularly.**

However, the committee's review of the literature and discussions with manufacturers indicate that, because of variabilities in the manufacturing process, the amount of lead from a CIVL can range from the equivalent of as few as 12,000 to as many as 35 million 40-grain, .22 caliber longrifle bullets compared with a total of 9 billion bullets produced each year. Further, there is the possibility that bullets from different CIVLs may be analytically indistinguishable. **Recommendation: Expert witnesses should define the range of CIVLs that could make up the source of analytically indistinguishable bullets because of variability in the bullet manufacturing process. The possible existence of coincidentally indistinguishable CIVLs should be acknowledged in the laboratory report and by the expert witness on direct examination.** The frequency with which coincidentally identical CIVLs occur is unknown.

Chapter 4 includes findings and recommendations about appropriate statements that can be made in laboratory reports or by expert witnesses based on the committee's findings on analytical methods and statistical procedures and its knowledge of the bullet manufacturing process, including the following:

- The available data do not support any statement that a crime bullet came from a particular box of ammunition. In particular, references to "boxes" of ammunition in any form should be avoided as misleading under Federal Rule of Evidence 403.
- Compositional analysis of bullet lead data alone also does not permit any definitive statement concerning the date of bullet manufacture.
- Detailed patterns of the distribution of ammunition are unknown, and as a result, experts should not testify as to the probability that the crime scene bullet came from the defendant. Geographic distribution data on bullets and ammunition are needed before such testimony can be given.

It is the conclusion of the committee that, in many cases, CABL is a reasonably accurate way of determining whether two bullets could have come from the same compositionally indistinguishable volume of lead. It may thus in appropriate cases provide additional evidence that ties a suspect to a crime, or in some cases evidence that tends to exonerate a suspect. CABL does not, however, have the unique specificity of techniques such as DNA typing to be used as standalone evidence. It is important that criminal justice professionals and juries understand the capabilities as well as the significant limitations of this forensic technique. The value and reliability of CABL will be enhanced if the recommendations set forth in this report are followed.

# 1

# Introduction

Compositional analysis of bullet lead (CABL) is chemical analysis of some (generally seven) of the elements found in lead alloy used to make bullets.[1] These elements may be present in lead ore but not completely removed in smelting, present in recycled lead used for bullet manufacture, or, as in the case of antimony, added to bullet lead to control such properties as hardness. In bullet manufacture, the concentrations of the elements in the lead alloy are specified only within broad ranges or below a maximum concentration, so given volumes of lead have differing elemental compositions.

The Federal Bureau of Investigation (FBI) has recognized and exploited that characteristic of bullet lead by using CABL. CABL allows bullets or bullet fragments found at a crime scene[2] to be compared with unused bullets found in the possession of a suspect.[3] Comparison is accomplished by using an analytical method that employs inductively coupled plasma-optical emission spectroscopy (ICP-OES).

ICP-OES is an instrumental method that is capable of determining the concentration of elements in solution. Each lead sample must be dissolved in an

---

[1] The same lead alloy is used to make bullet cores, lead projectiles that are swaged into a copper jacket before becoming part of a completed round of ammunition.

[2] Discussion of *bullets* and *bullet fragments* also includes shot from shotguns. Evidence considered to be *crime scene evidence* may be recovered at a crime scene or from a victim at a hospital or during an autopsy.

[3] It is possible that elemental analysis of the copper jacket from U.S.-produced, jacketed ammunition is less valuable than that of the bullet lead because of the tight industrial control of the purity of copper. Foreign manufacturers and some U.S. manufacturers may use alloys such as brass to form jackets; these alloys have not been studied as extensively as lead alloys.

acidic solution before analysis. The measurements of element concentrations obtained are compared to the measurements of element concentrations in a National Institute of Standards and Technology Standard Reference Material to determine the actual concentration of elements measured. If the concentrations of all seven elements in the bullet lead from a crime scene are determined by FBI examiners to statistically match the concentrations of the same seven elements in the bullet lead from a suspect, FBI examiners conclude that the bullets are "analytically indistinguishable." The results can be used by prosecutors as circumstantial evidence in a trial.

Some oppose the use of CABL. Questions have been raised as to the homogeneity of a source of lead, the uniqueness of a source of lead, the definition of a source of lead, the distribution of bullets and loaded ammunition, and the validity of specific statements made in court by expert witnesses.

- CABL assumes that a "source" of bullet lead is homogeneous. Opponents of CABL point to purported inadequate mixing of the lead melt in the manufacturing process as new materials are added, to the microscale separations that may occur during cooling of the bulk solid after the melt is poured, and to the migration of less-soluble elements to the interior of the solidifying lead as it cools after the melt is poured. If a source is not homogeneous, no bullet can be representative of the source.
- CABL also assumes that each lead source has a unique composition. Published data have shown that two lead sources prepared twelve years apart had compositions that were analytically indistinguishable[4] (Ref. 1).
- Analytically indistinguishable samples of bullet lead are said to come from the same source. There is some confusion about the definition of *source* and to which volume of lead in the manufacturing process it refers. The volume of lead affects the number of bullets that can be considered to come from one source.
- Although the major bullet manufacturers distribute their products nationally and even internationally, some regional distributors might receive and distribute many bullets from the same compositionally indistinguishable source. That would increase the probability of finding a match between a crime-scene bullet and a bullet in the possession of an innocent person.
- A wide variety of statements have been made in court by FBI examiners about the significance of CABL results. Some of these statements may have been exaggerated and may foster misinterpretation of the meaning of laboratory analyses.

The issues that have been raised by opponents to CABL are not trivial. To determine whether and how the use of CABL should be continued, the FBI

---

[4] A reanalysis of the samples may be needed because the published data lack specified assessments of reproducibility and repeatability.

wanted to address those issues and others to an independent, unbiased institution. Thus, the National Research Council (NRC) was called on to evaluate CABL scientifically, statistically, and legally. The questions in the statement of task accepted by the NRC with respect to CABL were as follows:

- *Analytical method.* Is the method analytically sound? What are the relative merits of the methods currently available? Is the selection of elements used as comparison parameters appropriate? Can additional useful information be gained by measurement of isotopic compositions?
- *Statistics for comparison.* Are the statistical tests used to compare two samples appropriate? Can known variations in compositions introduced in manufacturing processes be used to model specimen groupings and provide improved comparison criteria?
- *Interpretation issues.* What are the appropriate statements that can be made to assist the requester in interpreting the results of compositional bullet lead comparison, for both indistinguishable and distinguishable compositions? Can significance statements be modified to include effects of such factors as the analytical technique, manufacturing process, comparison criteria, specimen history, and legal requirements?

The Committee on Scientific Assessment of Bullet Lead Elemental Composition Comparison is composed of 14 experts in analytical chemistry, statistics, forensic science, metallurgy, and law. It met four times in Washington, D.C. The meetings allowed the committee to hear from experts in lead manufacturing, statistics, and use of CABL in court. At each meeting, the committee received presentations from FBI employees who research, use, or testify about CABL. The committee also used background information, such as scientific journal articles (both those provided to the committee by individuals outside the committee, and those found by the committee in its own search of relevant literature), published statistics on lead, court transcripts, and the expertise and experience of its members. Members of the committee visited the FBI Laboratory, Eldorado Cartridge Corporation/PMC, and the SHOT Show to gather data. The deliberations of the committee on the questions in the statement of task and on other related issues led to this report.

Chapter 2 addresses the analytical chemistry portion of CABL. It discusses the analysis of lead with ICP-OES and compares it with other, previously used instrumental methods and with potentially useful technology untested for this application. The elements that are measured with ICP-OES and compared to determine a match are also assessed. The chapter evaluates the entire written analytical protocol of the FBI and draws conclusions about the protocol's appropriateness and application.

Chapter 3 presents and critiques the statistical protocol used by the FBI for bullet matching. The chapter recommends alternate tests to be used in place of the FBI's current procedure.

The process of CABL culminates in its use as circumstantial evidence in court. The first half of Chapter 4 provides basic information about lead refining and bullet manufacturing to further an understanding of their significance in the interpretation of CABL data. It also offers some statistics on bullet production and the various volumes of liquid and solid lead that are eventually used to form bullets. Sections on the homogeneity of lead volumes and on the definition of *source* are integral to the committee's findings. The second half of the chapter introduces the admissibility of scientific evidence, relevance, and how CABL evidence has been used in trials. It discusses inconsistencies and changes in CABL-related testimony, laboratory reports, and printed handbooks and discusses the importance of these inconsistencies and changes. The chapter includes the rules governing pretrial discovery of reports and summaries of expert testimony, and the use of expert witnesses.

## REFERENCE

1. Randich, E.; Duerfeldt, W.; McLendon, W.; and Tobin, W. *Foren. Sci. Int.* **2002**, 127, 174–191.

# 2

# Compositional Analysis

The keystone of compositional analysis of bullet lead (CABL) is the analytical method. Before bullet matching, statistical analysis, or legal interpretation, the concentrations of elements in the bullet lead must be measured correctly. Any good analytical method relies on correct sample preparation, fitness of the instrument for the purpose, proper use of the instrumentation, and reliability. Proper documentation and transparency of the method are also necessary. Those topics are discussed in greater detail in this chapter.

## PREVIOUS INSTRUMENTAL METHODS

Historically, a number of instrumental methods have been used for the determination of elements in lead, including atomic absorption spectrometry (AAS),[1] neutron activation analysis (NAA)[2] spark source mass spectrometry (SSMS),[3] wavelength dispersive x-ray fluorescence (WDXRF) spectroscopy,[4]

---

[1]Brunnelle, R. L.; Hoffman, C. M.; and Snow, K. B., *JAOAC* **1970**, *53*, 470; Blacklock, E. C. and Sadler, P. A. *Foren. Sci. Int.* **1978**, 12, 109; Kramer, G. W. *Appl. Spec.* **1979**, 33, 468.; Krishnan, S. S. *Can. Soc. Foren. Sci. J.* **1972**, 6, 55; Gillespie, K. A. and Krishnan, S. S. *Can. Soc. Foren. Sci. J.* **1969**, 2, 95.

[2]Krishnan, **1972**; Gillespie and Krishnan, **1969**; Lukens, H. R.; Schlessinger, H. L.; Guinn, V. P.; and Hackleman, R. P. *US Atomic Energy Report GA-10401* **1970**; Lukens, H. R. and Guinn, V. P. *J. Foren. Sci.* **1971**, 16, 301; Guy, R. D. and Pate, B. D. *J. Radioanal. Chem.* **1973**, 15, 135.; Guinn, V. P. and Purcell, M. A. *J. Radioanal. Chem.* **1977**, 39, 85; Guinn, V. P. *J. Radioanal. Chem.* **1982**, 72, 645; Brandone, A. and Piancone, G. F. *J. Appl. Radiat. Isot.* **1984**, 35, 359.

[3]Haney, M. A. and Gallagher, J. F. *Anal. Chem.* **1975**, 47, 62.; Haney, M. A. and Gallagher, J. F. *J. Foren. Sci.* **1975**, 20, 484.

[4]Koons, R. D. *Spectroscopy* **1993**, 8(6), 16.

inductively coupled plasma-optical emission spectroscopy (ICP-OES),[5] and inductively coupled plasma-mass spectrometry (ICP-MS).[6] (The references cited in this paragraph are intended to document the historical progression of the analysis technique, and are not intended to represent the state of the art of current technology.)

Based on committee member's own expertise and knowledge of these techniques and familiarity with the recent literature, each of those instrumental methods has advantages and disadvantages. AAS is a single-element technique (one element at a time can be measured) that is limited in the overall number of elements that can be determined, although the elements of current interest for CABL can be determined. It also suffers from limited dynamic (working) range and is prone to interferences due to the sample matrix. NAA requires ready access to a nuclear reactor. SSMS has an advantage in that it requires minimal sample preparation; however, reliable quantitative analysis with SSMS is difficult. SSMS instrumentation also is not widely available. WDXRF spectroscopy suffers from inadequate limits of detection and has been used primarily for qualitative or semi-quantitative analysis.

ICP-MS has a sensitivity advantage over optical techniques, such as AAS and ICP-OES, and has a greater dynamic range than AAS. The major drawback of ICP-MS is that the lead sample matrix can suppress the element signals and can deposit on the sampling cone; this reduces ion throughput and yields erratic results.[7] That drawback can be avoided by precipitating the lead with sulfuric acid before ICP-MS analysis. However, the added precipitation step increases overall sample preparation time and lowers the precision and accuracy of the element measurements.

## INDUCTIVELY COUPLED PLASMA-OPTICAL EMISSION SPECTROPHOTOMETRY

The analytical characteristics of ICP-OES make it a useful technique for metal determinations.[8] A typical ICP-OES instrument has the following components:

---

[5]Peters, C. A.; Havekost, D. G.; and Koons, R. D. *Crime Lab. Digest* **1988**, 15, 33; Schmitt, T. J.; Walters, J. P.; and Wynn, D. A. *Appl. Spec.* **1989**, 43, 687; Peele, E. R.; Havekost, D. G.; Peters, C. A.; and Riley, J. P. USDOJ (ISBN 0-932115-12-8), 57, **1991**.

[6]Koons, R. D. *Spectroscopy*, **1993**, 8(6), 16; Suzuki, Y. and Marumo, Y. *Anal. Sci.* **1996**, 12, 129.

[7]Dufosse, T. and Touron, P. *Foren. Sci. Int.* **1998**, 91, 197; Jarvis, K. E.; Gray, J. L.; and Houk, R. S. *Inductively Coupled Plasma Mass Spectrometry*, Blackie & Son: London, 1992.

[8]Veale, N. P.; Olsen, L. K.; and Caruso, J. A. *Anal. Chem.* **1993**, 65 (13) 585A; Alcock, N. W. *Anal. Chem.* **1995**, 67 (12) 503R; *Methodology, Instrumentation, and Performance*, Boumans, P. W. J. M., Ed.; *Inductively Coupled Plasma Emission Spectroscopy Part 1*. John Wiley & Sons: New York, NY, 1987.

- Sample introduction system (nebulizer).
- Torch assembly.
- High-frequency generator.
- Transfer optics and spectrometer.
- Detector(s).
- Computer interface.

For analysis, samples generally are dissolved to form an aqueous solution of known weight and dilution. The solution is aspirated into the nebulizer, which transforms it into an aerosol. The aerosol then proceeds into the plasma, it is transformed into atoms and ions in the discharge, and the atoms (elements) are excited and emit light at characteristic wavelengths. The intensity of the light at the wavelengths associated with each element is proportional to that element's concentration.

The ICP-OES torch consists of three concentric tubes—known as the outer, middle, and inner tubes—usually made of fused silica. The torch is positioned in a coil of a radio-frequency generator. The support gas that flows through the middle annulus, argon, is seeded with free electrons that gain energy from the radio-frequency field. The energized electrons collide with the argon gas and form $Ar^+$ ions. Continued interaction of the electrons and ions with the radio-frequency field increases the energy of the particles and forms and sustains a plasma, a gas in which some fraction of the atoms are present in an ionized state. At the same time, the sample is swept through the inner loop by the carrier gas, also argon, and is introduced into the plasma, allowing the sample to become ionized and subsequently emit light.

Temperatures in the plasma are typically 6,000–10,000 K.[9] To prevent a possible short circuit and meltdown, the plasma must be insulated from the rest of the instrument. Insulation is achieved by the flow of the outer gas, typically argon or nitrogen, through the outer annulus of the torch. The outer gas sustains the plasma and stabilizes the plasma position.

Each element emits several specific wavelengths of light in the ultraviolet-visible spectrum that can be used for analysis. The selection of the optimal wavelength for a sample depends on a number of factors, such as the other elements present in the sample matrix. The light emitted by the atoms of an element must be converted to an electric signal that can be measured quantitatively. That is achieved by resolving the light with a diffraction grating and then using a solid-state diode array or other photoelectric detector to measure wavelength-specific intensity for each element emission line. The concentration of the elements in the sample is determined by comparing the intensity of the emission signals from the sample with that from a solution of a known concentration of the element (standard).

---

[9]Willard, H. H.; Merritt, Jr., L. L.; Dean, J. A.; Settle, Jr., F. A. *Instrumental Methods of Analysis, Seventh Ed.;* Wadsworth Publishing: Belmont, CA, 1988.

TABLE 2.1 Summary of Elemental Analysis Techniques

| Technique | Advantages | Disadvantages |
|---|---|---|
| AAS | Low detection limits | Few elements, time-consuming, matrix effects |
| NAA | Low detection limits | Few elements, requires access to reactor |
| SSMS | Low detection limits, multiple elements | Difficult quantification, surface-sensitive |
| WDXRF | Multiple elements, solid and liquid samples | Detection limits too high |
| ICP-MS | Low detection limits, multiple elements, isotope analysis | Matrix effects |
| ICP-OES | Low detection limits, multiple elements, limited spectral interferences, good stability, low matrix effects | Liquid samples only |

One of the main advantages of ICP-OES for elemental analysis is that it can be used to measure almost all the elements in the periodic table. The technique has a wide dynamic concentration range and can measure elements at trace to high concentrations. Detection limits for most elements are in the range of micrograms per liter to milligrams per liter. Another advantage of ICP-OES is that multielemental quantitative analysis can be carried out in a period as short as 1 min with a small amount of solution (0.5–1.0 mL). Those characteristics make ICP-OES a useful method for elemental analysis in forensic laboratories. ICP-OES is a technique that combines good quantitative multielement capability, wide linear dynamic ranges, good sensitivity, limited spectral and chemical interferences, low detection limits, and speed and ease of data handling and reporting with widespread (multiple-vendor) instrument availability and reasonable cost. Table 2.1 summarizes the advantages and disadvantages of ICP-OES and other elemental analysis techniques.

The Federal Bureau of Investigation (FBI) has been conducting bullet lead analysis for over 30 years. Initially, NAA was used to quantify three elements—antimony (Sb), copper (Cu), and arsenic (As)—in bullet lead. The FBI began to use ICP-OES in place of NAA in 1990, and over a period of several years expanded the list of elements to seven: arsenic, antimony, tin (Sn), copper, bismuth (Bi), silver (Ag), and cadmium (Cd).

## CURRENT FBI PROTOCOL

The "Principle and Scope" section of the current FBI procedure, *Comparative Elemental Analysis of Firearms Projectile Lead by ICP-OES*,[10] reads as follows:

[10]Peters, C. A. *Comparative Elemental Analysis of Firearms Projectile Lead by ICP-OES,* FBI Laboratory Chemistry Unit. Issue date: Oct. 11, 2002. *Unpublished* (2002).

> The concentrations of selected elements in the lead portion of bullets, shot pellets, and similar firearms projectiles serve to chemically characterize the source of lead. Some chemical elements present in these leads are intentionally specified and/or added by the ammunition manufacturer (e.g., antimony and arsenic). Other chemical elements typically found in these leads are present as unspecified contaminants (e.g., copper, tin, bismuth, and silver). Distinct and subtle differences in the concentrations of manufacturer controlled elements and uncontrolled trace elements provide a means of differentiating among the leads of different manufacturers, among the leads in individual manufacturers' product lines, and among specific batches of lead used in the same product line of a manufacturer.
>
> This procedure [ICP-OES] provides a method for determining and comparing the concentrations of seven elements: antimony, copper, arsenic, silver, tin, bismuth, and cadmium in the lead component of projectiles. Quantitative analysis is performed by dissolving the specimen and using the method of ICP-OES for measurement of individual element concentrations. Quantitation is achieved by comparison of specimens with a certified bullet lead reference standard ([National Institute of Standards and Technology Standard Reference Material] C2416).

The current FBI procedure is not documented in a complete and detailed format that would allow other laboratories skilled in the art to practice or even fully evaluate it. The "Principle and Scope" section of the documented procedure should be expanded to define the precision and accuracy of the method and the concentration ranges of all seven elements for which the method is applicable. Some precision data on the ICP-OES analytical method were presented in two FBI publications from 1988 and 1991[11] and are shown below in Tables 2.2 and 2.3. The published precision data, precision data from crime-scene and suspect bullet samples, and other, newer precision data more reflective of the current FBI CABL procedure should be included in the written protocol. The protocol should also describe how precision differs in the low, middle, and high ranges of each element's measurable concentrations.

The accuracy of the ICP-OES method was addressed by Schmitt et al.[12] and in an FBI publication from 1991.[13] Good statistical correlation was shown by Schmitt et al. between NAA and ICP-OES results for Cu and Sb.

The FBI's analytical procedure calls for three 60-mg samples (named a, b, and c at random) to be taken from each lead specimen through cutting. Representatives of the FBI informed the committee that each set of samples includes two calibration standards prepared from Standard Reference Material (SRM) C2416. Control samples derived from SRM C2416 (bullet lead), SRM C2415

[11] Peters et al., 1988; Peele et al. 1991.

[12] Schmidt et al., 1989.

[13] Peele et al., 1991.

TABLE 2.2 Within-Bullet Variability Measurements Based on ICP-OES

| Brand | Variability[a] | As | Sb | Sn | Cu | Bi | Ag |
|---|---|---|---|---|---|---|---|
| CCI | RSD, % | NA | 1.7 | NA | 1.7 | 3.8 | 1.9 |
| | Range, ppm | | 23,800–29,900 | | 97–381 | 56–180 | 18–69 |
| Federal | RSD, % | 3.7 | 1.5 | 2.5 | 1.5 | 6.7 | 2.3 |
| | Range, ppm | 1,127–1,645 | 25,700–29,000 | 1,100–2,880 | 233–329 | 30–91 | 14–19 |
| Remington | RSD, % | NA | 1.5 | NA | 1.5 | 3.4 | 1.8 |
| | Range, ppm | | 5,670–9,620 | | 62–962 | 67–365 | 21–118 |
| Winchester | RSD, % | NA | 1.9 | NA | 2.1 | 4.4 | 1.9 |
| | Range, ppm | | 2,360–6,650 | | 54–470 | 35–208 | 14–61 |

Note: RSD is relative standard deviation. NA indicates the data are not available because concentrations are too low to be accurately determined.
[a]Mean relative standard deviations of triplicate measurements of each bullet and the range in concentrations for all bullets of each brand examined in Peele et al. 1991; 10 bullets per brand were analyzed in triplicate.
Source: Table adapted from Peele et al., 1991.

TABLE 2.3 Precision of Analytical Results Based on ICP-OES

| Variability[a] | As | Sb | Sn | Cu | Bi | Ag |
|---|---|---|---|---|---|---|
| Range of concentrations of 50 bullets, µg/g | 1,000–1,900 | 2,500–6,800 | 1,400–2,600 | 71–483 | 53–221 | 14–56 |
| Mean RSD, % of triplicates | 3.4 | 1.7 | 3.5 | 2.0 | 5.3 | 2.7 |

[a]Mean relative standard deviations of triplicate measurements of 50 bullets.
Source: Taken from Peters et al., 1988.

(battery lead), and SRM C2417 (lead base alloy) are also included, as stated in the "Calibration and Control of Analytical Procedure" section of the FBI protocol.[14] All SRMs are lead-based alloys. The calibration and control samples are also divided into three sub-samples randomly labeled "a," "b," and "c."

The FBI's "Calibration and Control of Analytical Procedure" section lacks much of the information that is normally present in well-documented analytical protocols throughout the chemical industry. For example, standard FBI practice states that "a" calibration standards, "a" control samples, and all "a" series bullet lead sub-samples are run first, then the "b" series, and then the "c" series. This sequence is not described in the protocol. Although seemingly a minor detail,

[14] Peters, 2002.

this is of great importance because decisions are based on measurement precision, and factors that affect measurement precision need to be carefully controlled and documented.

The FBI's sample-digestion procedure for bullet lead evidence not only has evolved, but the committee learned, has not always been followed exactly. Once a single method is chosen, its viability should be ensured, and the procedure should be followed for every sample. It is most reliable if a universal procedure is used for *all* samples.

The "Decision Criteria" section of the FBI protocol describes the use of SRM C2416, SRM C2415, and SRM C2417 as *quality check samples.* Control values (limits) are given as means ± 2 standard deviations (SDs) for all seven elements. Most analytical laboratories use a formal control chart system. Such a system defines an average value of the measured variable, warning limits (means ± 2SD), and control limits (means ± 3SD), all based on historical data. If measured values are beyond the control limits, the process is considered to be out of control. Measured values outside the warning limits but within the control limits and values that are within the control limits but show trends (that is, movement in one direction or cyclical movement) are indicative of instrumental or procedural problems that should be fixed before the process becomes *out of control.*[15] A formalized control chart system would allow the FBI Laboratory to detect analytical problems early and keep the rate of false-positive matches low. Such a system is easily implemented with a software routine that translates collected data into standardized control charts.

The FBI (and perhaps other law-enforcement laboratories) has multiple examiners performing CABL and has employed many examiners over the lifetime of the technique. To ensure the validity of the CABL results, each examiner should be tested regularly for proficiency in carrying out the test. This proficiency testing should ensure the ability of the analyst to distinguish bullet fragments that are compositionally indistinguishable from fragments with similar but distinguishable compositions. As part of this testing, Gage R&R studies[16] should be carried out to assess the repeatability and reproducibility of the analysts involved in performing CABL. Proficiency testing is common in analytical laboratories and helps to ensure the overall quality of results. The proficiency tests are formalized and documented.[17]

---

[15] Vardeman, S. B. and Jobe, J. M. *Statistical Quality Assurance Methods for Engineers*, Wiley: New York, NY 1999.

[16] Vardeman and Jobe, 1999.

[17] One reviewer of this report suggested that the FBI laboratory should seek ISO certification to enhance its quality assurance and quality control. If the laboratory complies with the recommendations of the committee, its procedures should be compatible to the relevant sections of ISO 17025, the ISO standard most relevant to the laboratory. Because the FBI laboratory is not a commercial entity, the committee does not believe the time and expense involved in its obtaining full ISO certification is justified.

FBI representatives stated that distribution of the FBI's analytical protocol was tightly controlled until the document was requested by this committee. That controlled distribution was to ensure that only the newest version of the protocol was in use at any given time. But publication of the protocol and the research and data that support it in peer-reviewed journals or at a minimum publication of the protocol in other public venues would offer an opportunity for review and validation of the protocol. Publication options for the protocol include such a limited venue as *Forensic Science Communications* (on the FBI Web site), where the protocol could appear as a "Standards and Guidelines" article similar to "Standard Guide for Using Scanning Electron Microscopy/X-ray Spectrometry in Forensic Paint Examinations,"[18] and the *Federal Register,* which has a much broader distribution. Once the protocol is officially documented in the public domain, each FBI analyst should follow it without deviation.

## SELECTION OF COMPARISON ELEMENTS

The current FBI CABL method measures seven elements (As, Sb, Sn, Cu, Bi, Ag, and Cd). The selection of the elements has evolved, and it is unclear how their selection for comparison was made. The appropriateness of the elements selected depends on how discriminating the comparison of each element is in defining the composition of a volume of lead.

The FBI has published its assessment of the discriminating capabilities of individual elements in bullet lead comparisons.[19] The relative importance of the elements for discrimination between lead sources decreases in this order: Cu and As > Sb > Bi and Ag. Sn was not included in the appraisal, because it was not observed in the brands of ammunition used for the studies. Measurement of Cd was not added to the FBI's CABL procedure until 1995; therefore, Cd also was not included in the published studies.

A data set of elemental concentration measurements of bullet lead from 1,837 bullets compiled by the FBI was chosen as a basis for a statistical study of the discriminating ability of the seven elements. Information about the data set can be found in Chapter 3. Between-bullet standard deviations and correlations were calculated from the 1,837-bullet data set and demonstrated that correlation between the concentrations of some of the elements exist.

The variability in the 1,373-bullet subset can be characterized by using principal components analysis (PCA). PCA is a mathematical procedure that transforms a number of possibly correlated variables into a smaller number of noncorrelated variables called principal components. The most common use of PCA is dimension reduction: often, a fewer number of variables (defined as

---

[18]Unknown author, *Foren. Sci. Comm.*, 4(4), (2002).

[19] Peters et al., 1988; Peele et al. 1991.

TABLE 2.4 Assessment of Elemental Discriminating Ability Via Principal Components Analysis

| Elements | Percentage of Total Variation |
|---|---|
| Sb, Sn, Cd | 83.6 |
| Sb, Sn, Cd, As | 96.1 |
| Sb, Sn, Cd, As, Cu | 98.2 |
| Sb, Sn, Cd, As, Cu, Bi | 99.6 |
| Sb, Sn, Cd, As, Cu, Bi, Ag | 100 |

linear combinations of the original variables) contain a large proportion of variability of the entire data set. The first principal component accounts for as much of the variability in the data as possible, and each succeeding principal component accounts for as much of the remaining variability as possible. PCA was used here on the 1,373-bullet dataset (see Chapter 3, "Description of Data Sets") to compare the variability of the 1373 bullets when all 7 elemental measurements are used with the variability when all possible 3-, 4-, 5-, and 6-element subsets are used. By choosing the elements that contain most of the variability, one can minimize the false match probability. For complete details on how PCA was conducted, see Appendix H.

A summary of the results of PCA is given in Table 2.4. About 96% of the total variation was found with four elements (Sb, Sn, Cd, and As). The elements that contributed the least variation were Bi and Ag. The latter finding is consistent with the findings of the FBI and Randich.[20]

The results of PCA of the 1,373-bullet data set suggest that the FBI is obtaining the greatest amount of information and discrimination by measuring Sb, Sn, Cd, As, and Cu. Although little power to detect matches would be lost if Ag or Bi were dropped from the analytical procedure, using ICP-OES, no time or effort would be saved by measuring five rather than seven elements.

The committee considered whether analyzing additional elements would improve the predictive or matching power of CABL. Te and Se were focused on as the most promising candidates. Te in bullet lead has been quantified using ICP-MS.[21] However, Te, Se, and other elements that might be considered occur at ppm or sub-ppm levels, at or near the detection limit of the analytical technique. The precision of the measurement decreases quickly as measurements are taken near the detection limits of the instrument. As a result, the committee does not see analysis of additional elements as offering a significant improvement to the FBI's procedure.

[20]Randich, E.; Duerfeldt, W.; McLendon Sr., W.; and Tobin, W. *Foren. Sci. Int.* **2002,** 127, 174.
[21] Koons, 1993.

## INSTRUMENTAL METHODS FOR FURTHER STUDY

Some instrumental methods seem to hold promise for CABL. The most noteworthy are described below.

### Measurement of Lead Isotopic Compositions

The relative amounts of lead isotopes ($^{206}Pb$, $^{207}Pb$, and $^{208}Pb$) in different geographic regions can differ from 17% to 36%.[22] The reason for the variation in lead isotopic composition is the radioactive decay of thorium and uranium to lead by the following paths.[23]

$$^{238}U \rightarrow {}^{206}Pb$$
$$^{235}U \rightarrow {}^{207}Pb$$
$$^{232}Th \rightarrow {}^{208}Pb$$

If sufficient precision and mass resolution is available, ICP-MS may be able to distinguish the origins of lead on the basis of isotopic ratios. One early study used ICP-MS to distinguish lead sources (for example, paint, foundry ash, and soil) in pollution studies.[24] Although this technique does not appear to be particularly effective with domestically produced bullets that are made of lead from secondary smelters and thus may have a homogenized lead isotopic signature, some foreign bullets are made of lead from primary sources and could have characteristic lead isotopic signatures. The FBI may want to pursue research on this technique in the future.

### High Resolution Mass Spectrometry and Inductively Coupled Plasma-Mass Spectrometry

Initially, ICP-MS was dominated by low-resolution quadrupole-based instruments.[25] Although these instruments were sensitive and had lower limits of detection than ICP-OES, they were prone to interference problems, which limited their utility in lead isotopic analysis. The development of higher-resolution ICP-MS instruments—the first double-focusing ICP-MS commercial instruments appeared in the early 1990s[26]—may offer an improvement in the isotopic analysis of lead in bullets.

---

[22]Ault, W. U.; Senechai, R. E.; and Eriebach, W. E. *Environ. Sci. Tech.* **1970**, 4, 305; Brown, J. S. *Econ. Geol.* **1983,** 57, 673.

[23]Doe, B. R. *Lead Isotopes* Springer-Verlag: New York, NY, 1970.

[24]Hinners, T. A.; Heithmar, E. M.; Spittler, T. M.; and Henshaw, J. M. *Anal. Chem.* **1987**, 59, 2658.

[25]Houk, R. S. and Fassel, V. A. *Anal. Chem.* **1980**, 52, 2283; Houk, R. S. *Anal. Chem.* **1986**, 58, 97A.

[26]Stuewer, D. and Jakubowski, N. J. *Mass Spectrom.* **1998,** 33, 579.

One high-resolution MS approach for use in examining lead isotope ratios was reported by Andrasko et al.,[27] whose work demonstrated the ability of thermal ionization mass spectrometry (TIMS) to provide high-precision lead isotopic ratios for differentiating bullet samples. TIMS is the "standard" accepted method of isotopic ratio determination because of its potential precision. However, TIMS requires that the lead be separated from other elements before analysis because various mass-bias effects are generated during the ionization of lead from different matrices. This would be necessary whether the isotopic ratio determination was performed for lead or for any of the other trace elements in the bullet sample. The authors stated that this approach would be extremely difficult to implement on a routine basis.

More recently, a study was carried out with high-resolution ICP-MS based on a multi-collector (MC) system.[28] The use of multi-collectors is a key feature of TIMS that allows for simultaneous high-precision measurement of the isotopes of interest. The MC-ICP-MS instrument allows for the simultaneous measurement of the relevant lead isotopes, with the advantages of TIMS and the advantages of ICP-MS because it does not require the isolation of lead from other elements before analysis. The results showed that the MC-ICP-MS instrument had precision and accuracy that were about ten times better than those in a similar study of quadrupole ICP-MS.[29] Differences were observed with bullets obtained from economically isolated regions of the world, such as the former Soviet Union and South Africa. Although the study illustrated the possibility of differentiating between projectile lead in countries where a large amount of lead is recycled (such as the United States), the researchers were unable to utilize these analyses for determination of the lead deposit or source in such countries. Such a result would be expected whether the technique was used to measure the isotope ratio of the lead or of any of the trace elements in U.S.-manufactured bullets.

Suggested studies using the MC-ICP-MS approach would involve combining elemental analysis with the lead isotopic analysis in an attempt to increase the number of independent variables and improve the overall distinguishing ability of bullet lead analysis. The FBI should consider this for future study if foreign sources of bullet lead increase in the United States.

### Laser Ablation Inductively Coupled Plasma-Mass Spectrometry

Laser ablation (LA) coupled with ICP-MS has been increasingly studied over the last 5 years for the determination of elements in solid samples.[30] LA-

[27]Andrasko, J.; Koop, I.; Abrink, A.; and Skiold, T. J. *Foren. Sci.* **1993**, 38, 1161.

[28]Buttigieg, G.; Baker, M.; Ruiz, J.; and Denton, M.B. *Anal. Chem.*, in press.

[29]Dufosse and Touron, 1998.

[30]Winefordner, J. D.; Gornshukin, I. B.; Pappas, D.; Mateev, O. I.; and Smith, B.W. *J. Anal. At. Spectrom.* **2000**, 15, 1161; Tanaka, T.; Yamamoto, K.; Nomizu, T.; and Kawaguchi, H. *Anal. Sci.*

ICP-MS has a number of advantages for the analysis of solid samples, including minimal sample preparation, no loss of volatile elements, reduced contamination from reagents, and high sample throughput.

The main disadvantage of LA-ICP-MS is that its precision and accuracy are worse than those of ICP-MS with conventional pneumatic nebulization. Recently, several internal standard approaches were reported to improve overall accuracy and precision.[31] It may be advantageous to monitor future advancements of this method.

### High Performance Inductively Coupled Plasma-Optical Emission Spectroscopy

A method to improve measurement precision of ICP-OES by an order of magnitude or more was published in 1998; additional papers were published in 2000 and 2001.[32] The method is a ratio-based procedure that relies on the cancellation of correlated high-frequency noise in the instrument combined with a new way to reduce the effects of low-frequency signal drift. The drift-correction procedure models low-frequency drift in repeated measurements and corrects the data to a "drift-free" condition. Although the published method is quite involved, development of a simplified adaptation that could substantially improve the analytical precision of ICP-OES for bullet lead analysis might be possible. That could help to provide better discrimination between bullet compositions. The reliance on improved instrumental precision to improve discrimination assumes that this precision is a significant source of error in the overall measurement and evaluation procedure.

## FINDINGS AND RECOMMENDATIONS

**Finding:** The current analytical technology used by the FBI—inductively coupled plasma-optical emission spectroscopy (ICP-OES)—is appropriate and is currently the best available technology for the application.

**Recommendation:** The FBI Laboratory's analytical protocol should be revised to contain all details of the inductively coupled plasma-optical emission spec-

---

**1995**, 11, 967; Leach, J. J. Allen, L. A.; Aeschliman, D. B.; and Houk, R. S. *Anal. Chem.* **1990**, 71, 440; Gunther, D.; Hattendorf, B.; and Audetat, A. *J. Anal. At. Spectrom.* **2001**, 16, 1085; Mason, P. R. D. and Mank, A. J. G. *J. Anal. At. Spectrom.* **2001**, 16, 1381.

[31]Ohata, M.; Hiroyuki, Y.; Naimi, Y.; and Furuta, N. *Anal. Sci.* **2002**, 18, 1105.

[32]Salit, M. L. and Turk, G. C. *Anal. Chem.* **1998**, 70, 3184; Salit, M. L.; Vocke, R. D.; and Kelly, W. R. *Anal. Chem.* **2000**, 72, 3504; Salit, M. L.; Turk, G. C.; Lindstrom, A. P.; Butler, T. A.; Beck II, C. M.; and Norman, B. R. *Anal. Chem.* **2001**, 73, 4821.

troscopy (ICP-OES) procedure and to provide a better basis for the statistics of bullet comparison. Revisions should include:

(a) Determining and documenting the precision and accuracy of the ICP-OES method and the concentration range of all seven elements to which the method is applicable.

(b) Adding data on the correlation of older neutron activation analysis and more recent ICP-OES results and any additional data that address the accuracy or precision of the method.

(c) Writing and documenting the unwritten standard practice for the order of sample analysis.

(d) Modifying and validating the digestion procedure to assure that all of the alloying elements and impurities in all samples (soft lead and hard lead) are dissolved without loss.

(e) Using a more formal control-chart system to track trends in the procedure's variability.

(f) Defining a mechanism for validation and documentation of future changes.

**Recommendation:** The FBI should continue to measure the seven elements As, Sb, Sn, Cu, Bi, Ag, and Cd as stated in the current analytical protocol.

**Recommendation:** A formal and documented comprehensive proficiency test of each examiner needs to be developed by the FBI. This proficiency testing should ensure the ability of the analyst to distinguish bullet fragments that are compositionally indistinguishable from fragments with similar but analytically distinguishable composition. Testing could be internal or external (for example, conducted by the National Institute of Standards and Technology), and test results should be maintained and provided as appropriate. Proficiency should be tested regularly.

**Recommendation:** The FBI should publish the details of its CABL procedure and the research and data that support it in a peer-reviewed journal or at a minimum make its analytical protocol available through some other public venue.

**Recommendation:** Because an important source of measurement variation in quality-assurance environments may be the analyst who makes the actual measurements, measurement *repeatability* (consistency of measurements made by the same analyst) and *reproducibility* (consistency of measurements made by different analysts) need to be quantified through *Gage R & R studies.* Such studies should be conducted for the FBI comparison procedures.

**Recommendation:** The FBI's documented analytical protocol should be applied to *all* samples and should be followed by *all* examiners for *every* case.

**Recommendation:** The FBI should evaluate the potential gain from the use of high-performance inductively coupled plasma-optical emission spectroscopy because improvement in analytical precision may provide better discrimination.

# 3

# Statistical Analysis of Bullet Lead Data

## INTRODUCTION

Assume that one has acquired samples from two bullets, one from a crime scene (the CS bullet) and one from a weapon found with a potential suspect (the PS bullet). The manufacture of bullets is, to some extent, heterogeneous by manufacturer, and by manufacturer's production run within manufacturer. A CIVL, a "compositionally indistinguishable volume of lead"—which could be smaller than a production run (a "melt")—is an aggregate of bullet lead that can be considered to be homogeneous. That is, a CIVL is the largest volume of lead produced in one production run at one time for which measurements of elemental composition are analytically indistinguishable (within measurement error). The chemical composition of bullets produced from different CIVLs from various manufacturers can vary much more than the composition of those produced by the same manufacturer from a single CIVL. (See Chapter 4 for details on the manufacturing process for bullets.) The fundamental issue addressed here is how to determine from the chemical compositions of the PS and the CS bullets one of the following: (1) that there is a non-match—that the compositions of the CS and PS bullets are so disparate that it is unlikely that they came from the same CIVL, (2) that there is a match—that the compositions of the CS and PS bullets are so alike that it is unlikely that they came from different CIVLs, and (possibly) (3) that the compositions of the two bullets are neither so clearly disparate as to assert that they came from different CIVLs, nor so clearly similar to assert that they came from the same CIVL. Statistical methods are needed in this context for two important purposes: (a) to find ways of making these assertions based on the evidence so that the error rates—either the chance of falsely asserting a match, or the chance of falsely asserting a non-match, are both acceptably small, and (b) to estimate the size of these error rates for a given procedure, which need to be communicated along with the assertions of a match or a non-match so that the reliability of these assertions is understood.[1] Our general approach is to outline some of the possibilities and recommend specific statistical approaches for assessing matches and non-matches, leaving to others the selection of one or more critical values to separate cases 1), 2), and perhaps 3) above.[2]

Given the data on any two bullets (e.g., CS and PS bullets), one crucial objective of compositional analysis of bullet lead (CABL) is to provide information that bears on the question: "What is the probability that these two bullets were manufactured from the same CIVL?" While one cannot answer this question directly, CABL analysis can provide relevant evidence, the strength of that evidence depending on several factors.

First, as indicated in this chapter, we cannot guarantee uniqueness in the mean concentrations of all seven elements simultaneously. However, there is certainly variability between CIVLs given the characteristics of the manufacturing process and possible changes in the industry over time (e.g., very slight increases in silver concentrations over time). Since uniqueness cannot be assured, at best, we can address only the following modified question:

> "What is the probability that the CS and PS bullets would match given that they came from the same CIVL compared with the probability that they would match if they came from different CIVLs?"

The answer to this question depends on:

1. the number of bullets that can be manufactured from a CIVL,
2. the number of CIVLs that are analytically indistinguishable from a given CIVL (in particular, the CIVL from which the CS bullet was manufactured), and
3. the number of CIVLs that are not analytically indistinguishable from a given CIVL.

The answers to these three items will depend upon the type of bullet, the manufacturer, and perhaps the locale (i.e., more CIVLs may be more readily accessible to residents of a large metropolitan area than to those in a small urban town). A carefully designed sampling scheme may provide information from

[1] This chapter is concerned with the problem of assessing the match status of two bullets. If, on the other hand, a single CS bullet were compared with *K* PS bullets, the usual issues involving multiple comparisons arise. A simple method for using the results provided here to assess false match and false non-match probabilities is through use of Bonferroni's inequality. Using this method, if the PS bullets came from the same CIVL, an estimate of the probability that the CS bullet would match *at least one* of the PS bullets is bounded above by, but often very close to, *K* times the probability that the CS bullet would match a single PS bullet.

[2] The purposive selection of disparate bullets by those engaged in crimes could reduce the value of this technology for forensic use.

which estimates, and corresponding confidence intervals, for the probability in question can be obtained. No comprehensive information on this is currently available. Consequently, this chapter has given more attention to the only fully measurable component of variability in the problem, namely, the measurement error, and not to the other sources of variability (between-CIVL variability) which would be needed to estimate this probability.

Test statistics that measure the degree of closeness of the chemical compositions of two bullets are parameterized by critical values that define the specific ranges for the test statistics that determine which pairs of bullets are asserted to be matches and which are asserted to be non-matches. The error rates associated with false assertions of matches or non-matches are determined by these critical values. (These error rates we refer to here as the *operating characteristics* of a statistical test. The operating characteristics are often called the significance level or Type I error, and the power or Type II error.)

This chapter describes and critiques the statistical methods that the FBI currently uses, and proposes alternative methods that would be preferred for assessing the degree of consistency of two samples of bullet lead. In proposing improved methods, we will address the following issues:

1. General approaches to assessing the closeness of the measured chemical compositions of the PS and CS bullets,
2. Data sets that are currently available for understanding the characteristics of data on bullet lead composition,
3. Estimation of the standard deviation of measures of bullet lead composition, a crucial parameter in determining error rates, and
4. How to determine the false match and false non-match rates implied by different cut-off points (the critical values) for the statistical procedures advocated here to define ranges associated with matches, non-matches, and (possibly) an intermediate situation of no assertion of match status.

Before we address these four topics, we critique the procedures now used by the FBI. At the end, we will recommend statistical procedures for measuring the degree of consistency of two samples of bullet lead, leaving the critical values to be determined by those responsible for making the trade-offs involved.

## FBI's Statistical Procedures Currently in Use

The FBI currently uses the following three procedures to assert a "match," that is, that a CS bullet and a PS bullet have compositions that are sufficiently similar[3] for an FBI expert to assert that they were manufactured from CIVLs

[3]The term "analytically indistinguishable chemical composition" is used to describe two bullets that have compositions that are considered to match.

with the same chemical composition. First, the FBI collects three pieces from each bullet or bullet fragment (CS and PS), and nominally each piece is measured in triplicate. (These sample sizes are reduced when there is insufficient bullet lead to make three measurements on each of three samples.) Let us denote by $CS_i^k$ the $k^{th}$ measurement of the $i^{th}$ fragment of the crime scene bullet, and similarly for $PS_i^k$. Of late, this measurement is done using inductively coupled plasma-optical emission spectrophotometry (ICP-OES) on seven elements that are known to differ among bullets from different manufacturers and between different CIVLs from the same manufacturer. The seven elements are arsenic (As), antimony (Sb), tin (Sn), copper (Cu), bismuth (Bi), silver (Ag), and cadmium (Cd).[4]

The three replicates on each piece are averaged, and means, standard deviations, and ranges (minimum to maximum) for each element in each of the three pieces are calculated for all CS and PS bullets.[5] Specifically, the following are computed for each of the seven elements:

$$CS_i = \frac{CS_i^1 + CS_i^2 + CS_i^3}{3}$$

the average measurement for the $i^{th}$ piece from the CS bullet,

$$avg(CS) = \frac{CS_1 + CS_2 + CS_3}{3},$$

the overall average over the three pieces for the CS bullet,

$$sd(CS) = \sqrt{\frac{(CS_1 - avg(CS))^2 + (CS_2 - avg(CS))^2 + (CS_3 - avg(CS))^2}{2}},$$

the within-bullet standard deviation of the fragment means for the CS bullet—essentially the square root of the average squared difference between the average measurements for each of the three pieces and the overall average across pieces (the denominator uses 2 instead of 3 for a technical statistical reason),

$$range\ (CS) = \max(CS_1,CS_2,CS_3) - \min(CS_1,CS_2,CS_3),$$

the spread from highest to lowest of fragment means for the three pieces for the CS bullet.

The same statistics are computed for the *PS* bullet.

---

[4] As explained below, analyses in previous years measured only three to six elements, and in some cases, fewer than three pieces can be abstracted from a bullet or bullet fragment. However, in general, the following analysis will assume measurements on three pieces in triplicate for seven elements.

[5] Throughout this chapter, the triplicate measurements are ignored and the three averages are treated as the basic measurements. We have not found any analysis of the variability of measurements within a single sample; the FBI should conduct such an analysis as an estimate of pure measurement error, as distinct from variability within a single bullet. If the difference is trivial, use of the three fragments rather than the nine separate measurements is justified.

The overall mean, *avg(CS)*, is a measure of the concentration for a given element in a bullet. The overall mean could have differed: (1) had we used different fragments of the same bullet for measurement of the overall average, since even an individual bullet may not be completely homogeneous in its composition, and (2) because of the inherent variability of the measurement method. This variability in the overall mean can be estimated by the within-bullet standard deviation divided by $\sqrt{3}$ (since the mean is an average over 3 observations). Further, for normally distributed data, the variability in the overall mean can also be estimated by the range/3. Thus the standard deviation (divided by $\sqrt{3}$) and the range (divided by 3) can be used as approximate measures of the reliability of the sample mean concentration due to both of these sources of variation.

Since seven elements are used to measure the degree of similarity, there are seven different values of $CS_i$ and $PS_i$, and hence seven summary statistics for each bullet. To denote this we sometimes use the notation $CS_i$ $(As)$ to indicate the average for the $i^{th}$ bullet fragment for arsenic, for example, with similar notation for the other above statistics and the other elements.

### Assessment of Match Status

As stated above, in a standard application the FBI would measure each of these seven elements three times in each of three samples from the CS bullet and again from the PS bullet. The FBI presented to the committee three statistical approaches to judge whether the concentrations of these seven elements in the two bullets are sufficiently close to assert that they match, or are sufficiently different to assert a non-match. The three statistical procedures are referred to as: (1) 2-SD overlap, (2) range overlap, and (3) chaining. The crucial issues that the panel examined for the three statistical procedures are their operating characteristics, i.e, how often bullets from the same CIVL are identified as not matching, and how often bullets from different CIVLs are identified as matching. We describe each of these procedures in turn. Later, the probability of falsely asserting a match or a non-match is examined directly for the first two procedures, and indirectly for the last.

**2-SD Overlap** First, consider one of the seven elements, say arsenic. If the absolute value of the difference between the average compositions of arsenic for the CS bullet and the PS bullet is less than twice the sum of the standard deviations for the CS and the PS bullets, that is if $|avg(CS) - avg(PS)| < 2(sd(CS) + sd(PS))$, then the bullets are judged as matching for arsenic. Mathematically, this is the same criterion as having the 95 percent[6] confidence interval for the

[6]The 95 percent confidence interval for the *difference* of the two means, which is a more relevant construct for assessing match status, would utilize the square root of the variance of this difference, which is the square root of the sum of the two individual variances divided by the sample size for each mean (here, 3), not the sum of the standard deviations.

overall average arsenic concentration for the CS bullet overlap the corresponding 95 percent confidence interval for the PS bullet. This computation is repeated, in turn, for each of the seven elements. If the two bullets match using this criterion for all seven elements, the bullets are deemed a match; otherwise they are deemed a non-match.[7]

**Range Overlap** The procedure for range overlap is similar to that for the 2-standard deviation overlap, except that instead of determining whether 95 percent confidence intervals overlap, one determines whether the intervals defined by the minimum and maximum measurements overlap. Formally, the two bullets are considered as matching on, say, arsenic, if both $\max(CS_1,CS_2,CS_3) > \min(PS_1,PS_2,PS_3)$, and $\min(CS_1,CS_2,CS_3) < \max(PS_1,PS_2,PS_3)$. Again, if the two bullets match using this criterion for each of the seven elements, the bullets are deemed a match; otherwise they are deemed a non-match.

**Chaining** The description of chaining as presented in the FBI Laboratory document *Comparative Elemental Analysis of Firearms Projectile Lead by ICP-OES*, is included here as a footnote.[8] There are several different interpretations of this language that would lead to different statistical methods. We provide a

---

[7] The characterization of the 2-SD procedure here is equivalent to the standard description provided by the FBI. The equivalence can be seen as follows. Overlap is not occurring when either $avg(CS) + 2sd(CS) < avg(PS) - 2sd(PS)$ or $avg(PS) + 2sd(PS) < avg(CS) - 2sd(CS)$, which can be rewritten $avg(PS) - avg(CS) > 2(sd(CS) + sd(PS))$ or $avg(CS) - avg(PS) > 2(sd(CS) + sd(PS))$, which is equivalent to the single expression $|avg(CS) - avg(PS)| > 2(sd(CS) + sd(PS))$,

[8]a. CHARACTERIZATION OF THE CHEMICAL ELEMENT DISTRIBUTION IN THE KNOWN PROJECTILE LEAD POPULATION The mean element concentrations of the first and second specimens in the known material population are compared based upon twice the measurement uncertainties from their replicate analysis. If the uncertainties overlap in all elements, they are placed into a composition group; otherwise they are placed into separate groups. The next specimen is then compared to the first two specimens, and so on, in the same manner until all of the specimens in the known population are placed into compositional groups. Each specimen within a group is analytically indistinguishable for all significant elements measured from at least one other specimen in the group and is distinguishable in one or more elements from all the specimens in any other compositional group. (It should be noted that occasionally in groups containing more than two specimens, chaining occurs. That is, two specimens may be slightly separated from each other, but analytically indistinguishable from a third specimen, resulting in all three being included in the same compositional group.)

b. COMPARISON OF UNKNOWN SPECIMEN COMPOSITION(S) WITH THE COMPOSITION(S) OF THE KNOWN POPULATION(S): The mean element concentrations of each individual questioned specimen are compared with the element concentration distribution of each known population composition group. The concentration distribution is based on the mean element concentrations and twice the standard deviation of the results for the known population composition group. If all mean element concentrations of a questioned specimen overlap within the element concentration distribution of one of the known material population groups, that questioned specimen is described as being "analytically indistinguishable" from that particular known group population.

description here of a specific methodology that is consistent with the ambiguous FBI description. However, it is important that the FBI provide a rigorous definition of chaining so that it can be properly evaluated prior to use.

Chaining is defined for a situation in which one has a population of reference bullets. (Such a population should be collected through simple random sampling from the appropriate subpopulation of bullets relevant to a particular case, which to date has not been carried out, perhaps because an "appropriate" subpopulation would be very difficult to define, acquire, and test.) Chaining involves the formation of compositionally similar groups of bullets. This is done by first assuming that each bullet is distinct and forms its own initial "compositional group." One of these bullets from the reference population is selected.[9] This bullet is compared to each of the other bullets in the reference population to determine whether it is a match using the 2-SD overlap procedure.[10, 11] When the bullet is determined to match another bullet, their compositional groups are collapsed into a single compositional group. This process is repeated for the entire reference set. The remaining bullets are similarly compared to each other. In this way, the compositional groups grow larger and the number of such groups decreases.

This process is repeated, matching all of the bullets and groups of bullets to the other bullets and groups of bullets, until the entire reference population of bullets has been partitioned into compositional groups (some of which might still include just one bullet). Presumably, the intent is to join bullets into groups that have been produced from similar manufacturing processes. When the process is concluded, every bullet in any given compositional group matches at least one other bullet in that group, and no two bullets from different groups match.

The process to this point involves only the reference set. Once the compositional groups have been formed, let us denote the chemical composition (for one of the seven elements of interest) from the $k^{th}$ bullet in a given compositional group as $CG(k)$ $k = 1, \ldots, K$. Then the compositional group average and the compositional group standard deviations[12] are computed for this compositional group (assuming $K$ members) as follows, for each element:

$$avg(CG) = \frac{CG(1) + CG(2) + \ldots + CG(K)}{K},$$

[9]Assuming all bullets are ultimately compared to all other bullets, the order of selection of bullets is immaterial. Otherwise, the order can make a difference.

[10]The range overlap procedure could also be used.

[11]In the event that all three measurements for a bullet are identical, and hence the standard deviation is zero, the FBI specifies a minimum standard deviation and range for use in the computations.

[12]Note that the standard deviation of a compositional group with one member cannot be defined.

$$sd(CG) = \sqrt{\frac{(CG(1) - avg(CG))^2 + (CG(2) - avg(CG))^2 + ... + (CG(k) - avg(CG))^2}{(K-1)}},$$

Now, suppose that one has collected data for CS and PS bullets and one is interested in determining whether they match. If, for any compositional group, $|avg(CS) - avg(CG)| \leq 2sd(CG)$ for all seven elements, then the CS bullet is considered to be a match with that compositional group. (Note that the standard deviation of CS is not used.) If using the analogous computation, the PS bullet is also found to be a match with the same compositional group, then the CS and the PS bullets are considered to be a match.

This description leaves some details of implementation unclear. (Note that the 7-dimensional shapes of the compositional groups may have odd features; one could even be completely enclosed in another.) First, since $sd(CG)$ is undefined for groups of size one, it is not clear how to test whether the CS of PS bullets matches a compositional group of one member. Second, it is not clear what happens if the CS or the PS bullet matches more than one compositional group. Third, it is not clear what happens when neither the CS nor the PS bullets match any compositional groups.

An important feature of chaining is that in forming the compositional groups with the reference population, if bullet A matches bullet B, and similarly if bullet B matches bullet C, bullet A may not match bullet C. (An example of the variety of bullets that can be matched is seen in Figure 3.1.) One could construct examples (which the panel has done using data provided by the FBI) in which large chains could be created and include bullets that have little compositionally in common with others in the same group. Further, a reference bullet with a large standard deviation across all seven chemical compositions has the potential of matching many other bullets. Having such a bullet in a compositional group could cause much of the non-transitivity[13] just described.

Also, as more bullets are added to the reference set, any compositional groups that have been formed up to that point in the process may be merged if individual bullets in those compositional groups match. This merging may reduce the ability of the groups to separate new bullets into distinct groups. In an extreme case, one can imagine situations in which the whole reference set forms a single compositional group. The extent to which distinctly dissimilar bullets are assigned to the same compositional group in practice is not known, but clearly chaining can increase the rate of falsely asserting that two bullets match in comparison to the use of the 2-SD and range overlap procedures.

The predominant criticisms of all three of these procedures are that (1) the

[13] Non-transitivity is where A matches B, and B matches C, but A does not match C.

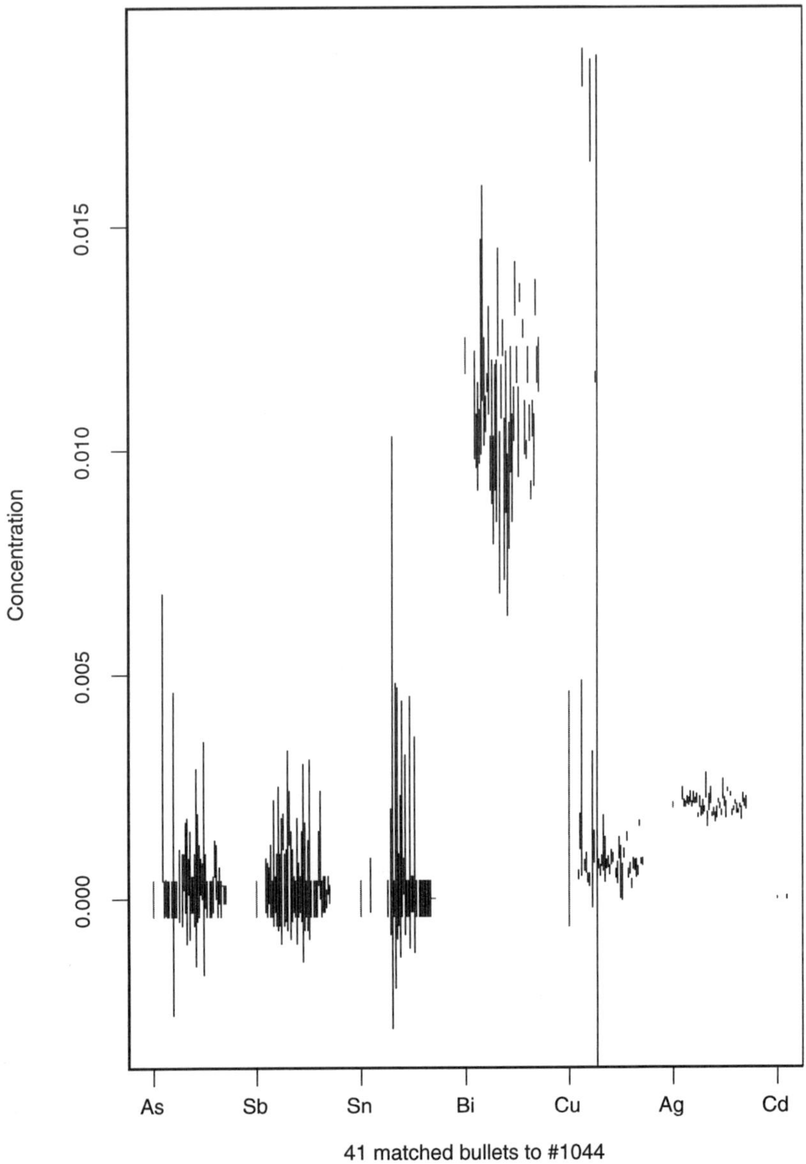


FIGURE 3.1 Illustration of chaining shows the 2-SD interval for bullet 1044 (selected at random) as first line in each set of elements, followed by the 2-SD interval for each of 41 bullets whose 2-SD intervals overlap with that of bullet 1044.

error rates for false matching and false non-matching are not known, even if one were to assume that the measured concentrations are normally distributed, and (2) these procedures are less efficient, again assuming (log) normally distributed data, in using the bullet lead data to make inferences about matching, than competing procedures that will be proposed for use below.

## Distance Functions

In trying to determine whether two bullets came from the same CIVL, one uses the "distance" between the measurements as the starting point. For a single element, the distance may be taken as the difference between the values obtained in the laboratory. Because that difference depends, at least in part, on the degree of natural variation in the measurements, it should be adjusted by expressing it in terms of a standard unit, the standard deviation of the measurement. The standard deviation is not known, but can be estimated from either the present data set or data collected in the past. The form of the distance function is then:

$$|\bar{x} - \bar{y}| / s,$$

where $s$ is the estimate of the standard deviation.

The situation is more complicated when there are measurements on two separate elements in the bullets, though the basic concept is the same. One needs the two-dimensional distance between the measurements and the natural variability of that distance, which depends on the standard deviations of measurements of the two elements, and also on the correlation between them. To illustrate in simple terms, if one is perfectly correlated (or perfectly negatively correlated) with the other, the second conveys no new information, and vice versa. If one measurement is independent of the other, distance measures can treat each distance separately. In intermediate cases, the analyst needs to understand how the correlation between measurements affects the assessment of distance. One possible distance function is the largest difference for either of the two elements. A second distance function is to add the differences across elements; this is equivalent to saying that the difference between two street addresses when the streets are on a grid is the sum of the north-south difference plus the east-west difference. A third is to take the distance "as the crow flies," or as one might measure it in a straight line on a map. This last definition of distance is in accord with many of our uses and ideas about distance, but might not be appropriate for estimates of (say) the time needed to walk from one place to another along the sidewalks. Other distance functions could also be defined. Again, we only care about distance and not direction, and for mathematical convenience we often work with the square of the distance function.

The above extends to three dimensions: One needs an appropriate function of the standard deviations and correlations among the measurements, as well as a

**Technical details on the $T^2$ test**

For any number $d$ of dimensions (including one, two, three, or seven)

$$T^2 = n(X - Y)'S^{-1}(X - Y)$$

where $X$ is a vector of seven average measured concentrations on the CS bullet, $Y$ is a vector of seven average measured concentrations on the PS bullet, $'$ denotes matrix transposition, $n$ = number of measurements in each sample mean (here, $n = 3$) and $S^{-1}$ = inverse of the 7 by 7 matrix of estimated variances and covariances.

Under the assumptions that

- the measurements are normally distributed (if lognormal, then the logarithms of the measurements are normally distributed),
- the matrix of variances and covariances is well-estimated, using $\nu$ degrees of freedom (for example, $\nu = 200$, if three measurements are made on each of 100 bullets and the variances and covariances within each set of three measurements are pooled across the 100 bullets),
- and the difference in the means of $X$ and $Y$ is $\delta = (\delta_1, \ldots, \delta_7)$and the standard deviation of $X$ equals the standard deviation of $Y$ equals $(\sigma_1, \sigma_2 \ldots, \sigma_7)$

then: $[(\nu - 6)/7\nu]T^2$ should not exceed a critical value determined by the noncentral F distribution with $p$ and $\nu$ degrees of freedom and noncentrality parameter, which is a function of $\delta$, $\sigma$, and $S^{-1}$.

When $\nu = 400$ degrees of freedom, and using the correlation matrix estimated from the data from one of the manufacturers of bullet lead (which measured six of the seven elements with ICP-OES; see Appendix F), and assuming that the measurement uncertainty on Cd is 5 percent and is uncorrelated with the others, the choice of the following critical values will provide a procedure with a false match

specific way to define difference (e.g., if the measurements define two opposite corners of a box, one could use the largest single dimension of the box, the sum of the sides of the box, the distance in a straight line from one corner to the other, or some other function of the dimensions). Again, the distance is easier to use if it is squared.

These concepts extend directly to more than three measurements, though the physical realities are harder to picture. A specific, squared distance function, generally known as Hotelling's $T^2$, is generally preferred over other ways to define the difference between sets of measurements because it summarizes the information on all of the elements measured and provides a simple statistic that has small error under common conditions for assessing, in this application, whether the two bullets came from the same CIVL.

rate, due to measurement error, of no more than 0.0004 (1 in 2,500—which is equivalent to the current asserted false match rate for 2-SD overlap): assert a match when $T^2$ is less than 1.9, assuming $\delta / \sigma = 1$ for each element, and assert a match when $T^2$ is less than 6.0, assuming $\delta / \sigma = 1.5$ for each element, where $\delta$ is the true difference between each elemental concentration and $\sigma$ is the true within-bullet standard deviation, i.e., the elemental measurement error assuming no within-bullet heterogeneity.

The critical value 1.9 requires that several assumptions be at least approximately true. There is the assumption of (log) normality of the concentration measurements. The use of $T^2$ is sensitive to the estimation of the inverse of the covariance matrix, and $T^2$ assumes that the differences in element concentrations are spread out across all seven elements fairly equally rather than concentrated in only one or two elements. (The latter can be seen from the fact that, if the measurement errors were independent, $T^2/7$ reduces to the average of squared two-sample *t* statistics for the $p = 7$ separate elements, so one moderately large difference will be spread out across the seven dimensions, causing $[(\nu - 6) / \nu]T^2/7$ to be small and thus to declare a match when the bullets differ quite substantially in one element.)

Unfortunately, the validity of Hotelling's $T^2$ test in the face of departures from those assumptions is not well understood. For example, the limit 1.9 is based on an estimated covariance matrix from one set of 200 bullets from one study conducted in 1991 (given in Appendix F), and the inferences from it may not apply to the current measurement procedure or to the bullets now produced. Many more studies would be needed to assess the reliability of $T^2$ in this application, including examination of the differences typically seen between bullet concentrations, the precision of estimates of the variances and covariances between measurement errors, and sensitivity to the assumption of (log) normality.

Source: *Multivariate Statistics Methods*, 2nd edition, Donald F. Morrison, McGraw-Hill Book Co., New York, NY, 1976.

## Statistical Power

Conclusions drawn from a statistical analysis of the distance between two sets of measurements can be wrong in either of two ways. In the case of bullet lead, if the bullets are in fact from the same CIVL, a conclusion that they are from CIVLs with different means is wrong. Conversely, if the means of the CIVL are not the same, a decision that they are the same is also an error. The latter error may occur when the two bullets from different CIVLs have different compositions but are determined to be analytically indistinguishable due to the allowance for measurement error, or when the two CIVLs in question have by coincidence the same chemical composition. The two kinds of error occur in incompatible situations, one where there is no difference and one where there is. Difficulties arise because we do not know which situation holds, so we must protect ourselves as well as possible against both types of error.

"Power" is a technical term for the probability that a null hypothesis will be rejected at a given significance level given that an alternative hypothesis is in effect. Generally, we want the power of a statistical test to be high for detecting a difference when one exists. The probabilities of the two kinds of error, the significance level—the probability of rejecting the null hypothesis when it is true, and one minus the power—the probability of failing to reject the null hypothesis when it is false, can be partly controlled through the use of efficient statistical procedures, but it is not possible to control both separately. For any given set of data, as one error is decreased, the other inevitably increases. Thus one must try to find an appropriate balance between the two types of error, which is done through the choice of critical values.

For a univariate test of the type described here, critical values are often set so that there is a 5 percent chance of asserting a non-match when the bullets actually match, i.e., 5 percent is the false non-match rate. This use of 5 percent is entirely arbitrary, and is justified by many decades of productive use in scientific studies in which data are generally fairly extensive and of good quality, and an unexpected observation can be investigated to determine whether it was a statistical fluke or represents some real, unexpected phenomenon.

If one examines a situation in which the difference between two bullets is very nearly, but not equal to zero, the probability of asserting a non-match for what are in fact non-matching bullets will remain close to 5 percent. However, as the difference between the bullets grows, the probability of asserting a non-match will grow to virtually 100 percent.

In the application of hypothesis testing to the issue at hand, there is an advantage in using as the null hypothesis, rather than the standard null hypothesis that the means for the two bullets are equal, the null hypothesis that the two means differ by greater than the measurement uncertainty. This has the advantage of giving priority, under the usual protocol, to the setting of the size of the test, which is then the false match probability, rather than using the standard null hypothesis, which would give priority to the false non-match probability. However, in the following we adopt a symmetric approach to the two types of errors, suggesting that both be estimated and that they be chosen to have socially acceptable levels of error.

## DESCRIPTION OF DATA SETS

This section describes three data sets made available to the committee that were used to help understand the distributional properties of data on the composition of bullet lead. These three datasets are denoted here as the "800-bullet data set," the "1837-bullet data set," and the "Randich et al. data set." We describe each of these data sets in turn.

TABLE 3.1 Number of Cases Having *b* Bullets in the 1837-Bullet Data Set

| *b* = no. bullets | 1 | 2 | 3 | 4 | 5 | 6 | 7 | 8 | 9 | 10 | 11 | 14 | 21 |
|---|---|---|---|---|---|---|---|---|---|---|---|---|---|
| No. cases | 578 | 283 | 93 | 48 | 24 | 10 | 7 | 1 | 1 | 2 | 1 | 1 | 1 |

**800-bullet Data Set**[14] This data set contains triplicate measurements on 50 bullets in 16 boxes—four boxes from each of four major manufacturers (CCI, Federal, Remington, and Winchester) measured as part of a study conducted by Peele et al. (1991). For each of the four manufacturers, antimony (Sb), copper (Cu), and arsenic (As) were measured with neutron activation analysis (NAA), and antimony (Sb), copper (Cu), bismuth (Bi), and silver (Ag) were measured with ICP-OES. In addition, for the bullets manufactured by Federal, arsenic (As) and tin (Sn) were measured using both NAA and ICP-OES. In total, this data set provided measurements on 800 bullets with Sb, Cu, Bi, and Ag, and 200 bullets with measurements on these and on As and Sn. This 800-bullet data set provides individual measurements on three bullet lead samples which permits calculation of within-bullet means, standard deviations, and correlations for six of the seven elements measured with ICP-OES (As, Sb, Sn, Bi, Cu, and Ag). In our analyses, the data are log-transformed. Although the data refer to different sets of bullets depending on the element examined, and have some possible outliers and multimodality, they are the only source of information on within-bullet correlations that the committee has been able to find.

**1,837-bullet Data Set**[15] The bullets in this data set were extracted from a historical file of more than 71,000 bullets analyzed by the FBI laboratory. The 1,837 bullets were selected from the larger set so as to include at least one bullet from each individual case that was determined, by the FBI chemists, to be distinct from the other bullets in the case.[16] (This determination involved the bullet caliber, style, and nominal alloy class.) Bullets from 1,005 different cases that occurred between 1989 and 2002 are included. The distribution of number of bullets per case (of the bullets selected for the data set) is given in Table 3.1.

[14] The 800-bullet data set was provided by the FBI in an e-mail from Robert D. Koons to Jennifer J. Jackiw, dated February 24, 2003. Details on the origin of the data set were provided to the panel by R.D. Koons in a personal communication on May 12, 2003. For additional details, see Peele et al. (1991).

[15] The 1,837-bullet data set was provided by the FBI; received by the committee on May 12, 2003.

[16] According to the notes that accompanied the data file, the bullets in it were selected to include one or more bullets that were determined to come from melts that were different from the other bullets in the data set; a few are research samples "not associated with any particular case," and a few "were taken from the ammunition collection (again, not associated with a particular case)."

While all bullets in the 1,837-bullet data set were to be measured three times using three fragments from each bullet, only the averages and standard deviations of the (unlogged) measurements are available. As a result, estimation of the measurement uncertainty (relative standard deviation within bullets) could only be estimated with bias. Further, a few of the specified measurements were not recorded, and only 854 bullets had all seven elements measured. Also, due to the way in which these bullets were selected, they do not represent a random sample of bullets from the population of bullets analyzed by the laboratory. The selection likely produced a dataset whose variability between bullets is higher than would be seen in the original complete data set, and is presumably higher than in the population of all manufactured bullets. This data set was useful for providing the committee with approximate levels of concentrations of elements that might be observed in bullet lead.[17]

A particular feature of this data set is that the data on Cd are highly discrete: 857 measurements are available of which 285 were reported as 0, 384 of the 857 had Cd concentrations equal to one of six measurements (10, 20, 30, 40, 50, or 60 ppm), and the remaining 188 of the 857 available measurements were spread out from 70 to 47,880 ppm. (The discreteness of the measurements below 70 ppm stem from the precision of the measurement, which is limited to one significant digit due to dilutions in the analytical process.) Obviously, the assumption of log-normality is not fully supportable for this element. We at times focus our attention here on the 854-bullet subset with complete measurements, but also utilize the entire data set for additional computations.

**Randich et al. (2002)** These data come from Table 1 in an article by Randich et al. (2002). Six elements (all but Cd) were measured for three samples from each of 28 lead castings. The three samples were selected from the beginning, middle, and end of each lot. This data set was used to compare the degree of homogeneity of the lead composition in a lot to that between lots.

Each of these three data sets has advantages but also important limitations for use in modeling the performance of various statistical procedures to match bullet lead composition, especially with respect to determining the chances of asserting a false match or a false non-match. The 800-bullet data set has somewhat limited utility since it has data from only four manufacturers, though they are the major manufacturers in the United States and account for the majority of bullets made domestically. If those manufacturers are in any way unrepresentative of the remaining manufacturers, or if the CIVLs analyzed are for some reason not representative of what that manufacturer distributes, the data can tell us little about the composition of bullets from other manufacturers or CIVLs. However, the 800-bullet data set does provide important information on within-

[17] See Appendix F for details on within-bullet correlations.

bullet measurement variability and the correlations between various pairs of different elemental composition measurements within a bullet. The analyses in Carriquiry et al. (2002) and Appendix F show that it is reasonable to assume that these estimated parameters are not strongly heterogeneous across manufacturer. This type of analysis is important and should be continued.

The 1,837-bullet data set and the subset we have used are affected by three main problems. First, since the bullets were selected so that the FBI was relatively certain that the bullets came from different melts, the variability represented in the data set is likely to be greater than one would anticipate for bullets selected at random from different melts (which we discuss below). Therefore, two bullets chosen from different CIVLs, as represented in this data set, might coincidentally match less often than one would observe in practice when bullets come from different melts. The extent of any such bias is unknown. In addition, there is a substantial amount of missing data (some elements not measured), which sometimes forces one to restrict one's attention to the 854 bullets for which measurements of the concentration of all seven elements are available. Finally, the panel was given the means, but not the three separate measurements (averaged over triplicates), on each bullet so that within-bullet correlations of the compositions of different elements cannot be computed.

The data of Randich et al. (2002) provide useful information on the relative degree of homogeneity in a lot in comparison to that between lots, and hence on the degree of variation within a lot in comparison to that between lots. However, as in the 800-bullet data set, these data are not representative of the remaining manufacturers, and one element, Cd, was not measured. Inhomogeneity implies that one lot may contain two or more CIVLs.

In summary, we will concentrate much of our analysis on the 1,837-bullet data set, understanding that it likely has bullets that are less alike than one would expect to see in practice. The 1,837-bullet data set was used primarily to validate the assumption of lognormality in the bullet means, and to estimate within-bullet standard deviations. However, the 1,837-bullet data set, while providing useful information, cannot be used for unbiased inferences concerning the general population of bullets, or for providing unbiased estimates of the error rates for a test procedure using as inputs bullet pairs sampled at random from the general population of bullets. The Randich and the 800-bullet data sets were utilized to address specific issues and to help confirm the findings from the 1,837 (854) bullet data set.

## Properties of Data on Lead Composition

### Univariate Properties

The data on composition of each of the seven elements generally, but not uniformly, appear to have a roughly *lognormal* distribution. (See Figures 3.2, 3.3, 3.4, and 3.5 for histograms on elemental composition.) That is, the data are

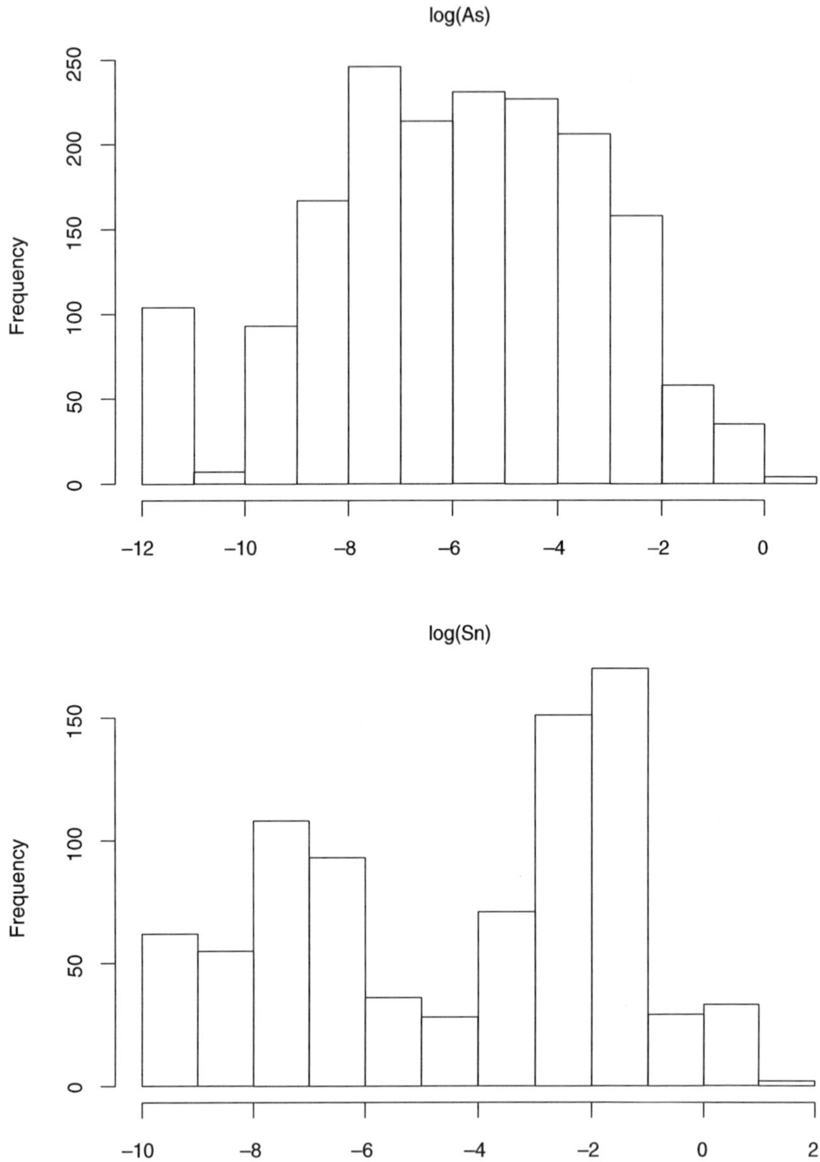


FIGURE 3.2 Histograms of mean concentrations (ppm) in bullets from 1,837-bullet data set: (a) log(As mean concentrations); (b) log(Sn mean concentrations).

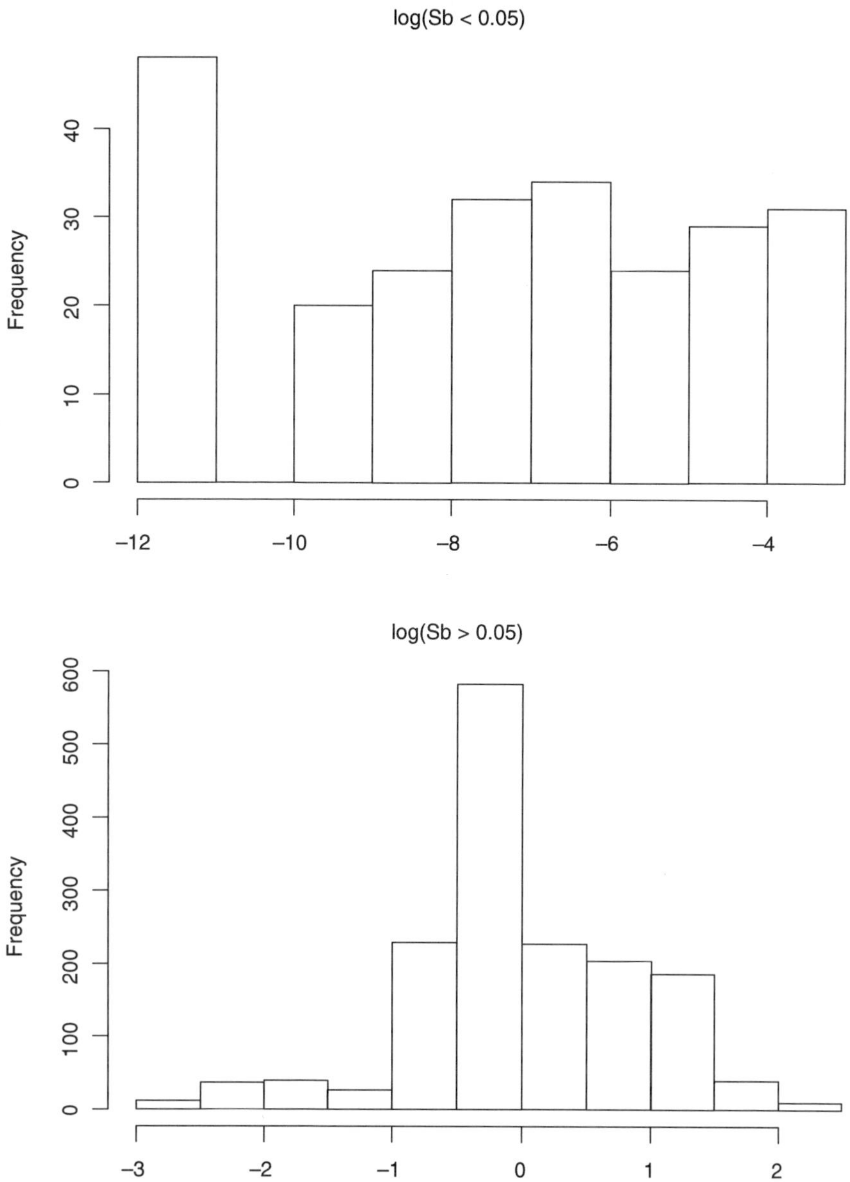


FIGURE 3.3 Histograms of Sb mean concentrations (ppm) in bullets from 1,837-bullet data set: (a) log(Sb mean concentrations less than 0.05); (b) log(Sb mean concentrations greater than 0.05).

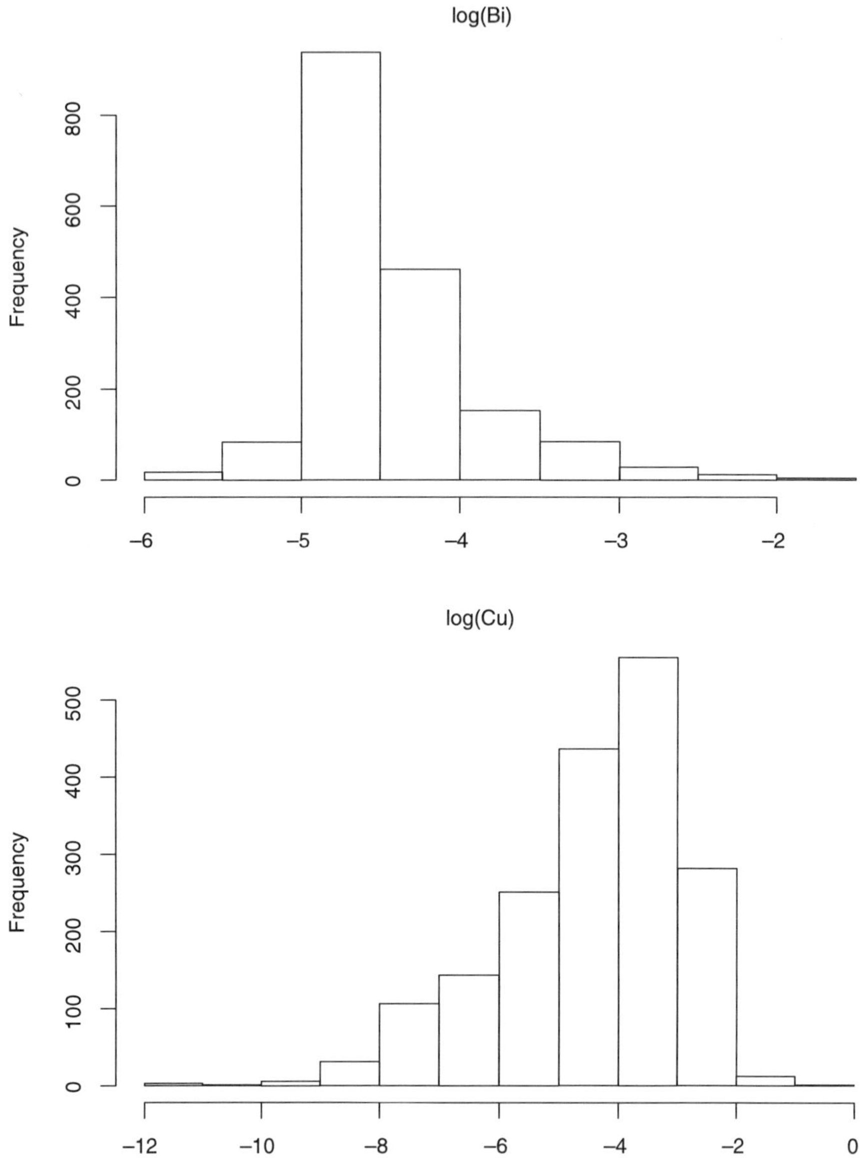


FIGURE 3.4 Histograms of Bi and Cu mean concentrations (ppm) in bullets from 1,837-bullet data set: (a) log(Bi mean concentrations); (b) log(Cu mean concentrations).

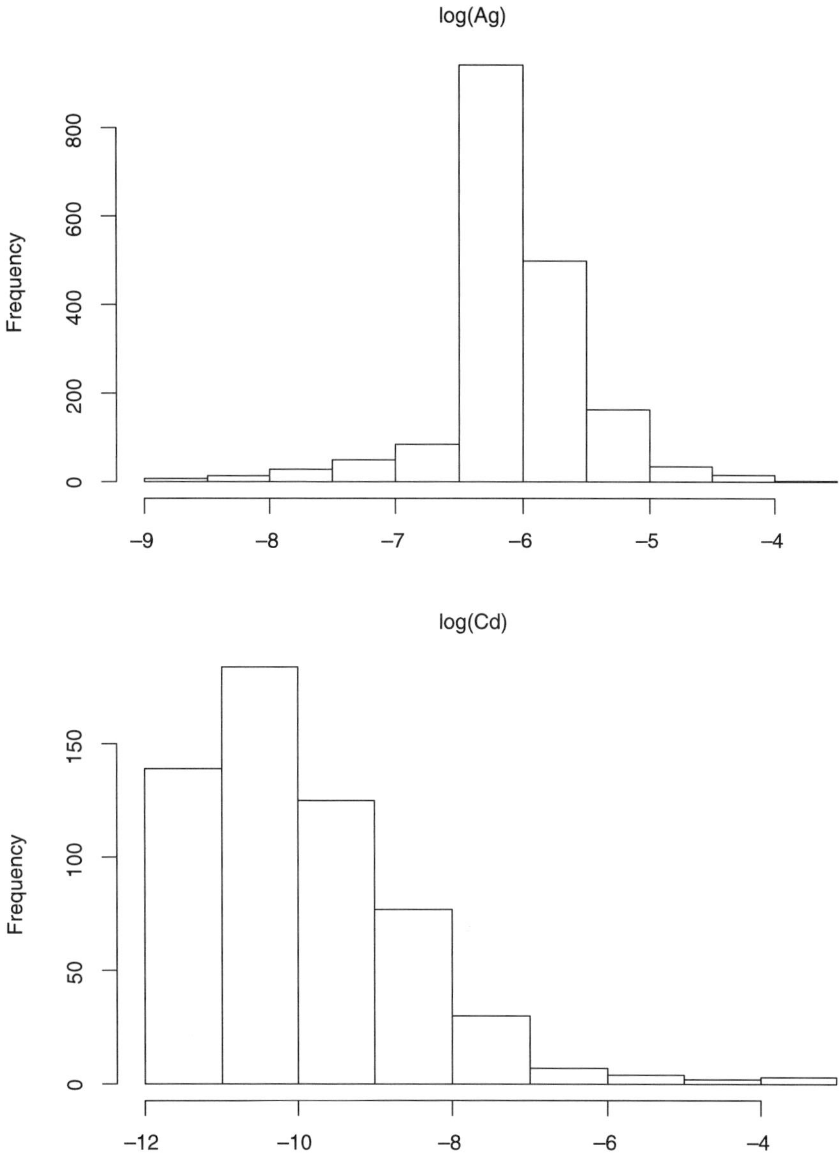


FIGURE 3.5 Histograms of Ag and Cd mean concentrations (ppm) in bullets from 1,837-bullet data set: (a) log(Ag mean concentrations); (b) log(Cd mean concentrations).

distributed so that their logarithms have an approximately normal distribution. The lognormal distribution is asymmetric, with a longer right tail to the distribution. The more familiar normal distribution that results from taking logarithms has the advantage that many classical statistical procedures are designed for, and thus perform optimally on, data with this distribution.

The 1,837-bullet data set revealed that the observed within-bullet standard deviations (as defined above for CS and PS) are roughly proportional to the measured bullet averages. In contrast, data from the normal distribution have the same variance, regardless of their actual value. For this reason, it is common in this context to refer to the *relative standard deviation* (RSD), which is defined as 100(*stdev / mean*). Taking logarithms greatly reduces this dependence of variability on level, which again results in a data set better suited to the application of many classical statistical procedures. Fortunately, standard deviations computed using data that have been log-transformed are very close approximations to the RSD, and in the following, we will equate RSD on the untransformed scale with the standard deviation on the logarithmic scale. (For details, see Appendix E.)

However, the data for the seven elements are not all lognormal, or even mixtures of lognormal data or other simple generalizations. We have already mentioned the discrete nature of the data for cadmium. In addition, the 1,837-bullet data set suggests that, for the elements Sn and Sb, the distributions of bullet lead composition either are bimodal, or are mixtures of unimodal distributions. Further, some extremely large within-bullet standard deviations for copper and tin are not consistent with the lognormal assumption, as discussed below. This is likely due either to a small number of outlying values that are the result of measurement problems, or to a distribution that has a much longer right-side tail than the lognormal. (Carriquiry et al. (2002) utilize the assumption of mixtures of lognormal distributions in their analysis of the 800-bullet data set.)

A final matter is that the data show evidence of changes over time in silver concentration in bullet lead. Most of the analysis carried out and techniques proposed for use assume that the data are from single, stable distributions of bullet-lead concentrations. Variation in concentrations over time could have a substantial impact on the operating characteristics of the statistical tests discussed here (likely making them more effective due to the added difference between bullets manufacturer at different times), resulting in estimated error rates that are higher than the true rates. However, the dynamics might be broader, e.g., making one of the seven elements less important to include in the assessment, or possibly making it useful to add other elements. This can be partially addressed by using a standard data set that was generated from bullets made at about the same time as the bullet in question. Unfortunately, one does not in general know when a CS bullet was made. This issue needs to be further examined, but one immediate step to take is to regularly measure and track element concentrations and compute within-bullet standard deviations and correlations to ensure the stability of the measurements and the measurement process. A standard statistical construct, the control chart, can be used for this purpose. (See Vardeman and Jobe (1999) for details.)

## Within-Bullet Standard Deviations and Correlations

From the 800-bullet data set of the average measurements on the logarithmic scale for each bullet fragment, one can estimate the within-bullet standard deviation for each element and the within-bullet correlations between elements. (We report results from the log-transformed data, but results using the untransformed measurements were similar).

Let us refer to the chemical composition of the $j^{th}$ fragment of the $i^{th}$ bullet from the 800-bullet data set on the log scale as $As_i^j$, and the average (log) measurement over the three fragments as $As_i^+$, where As stands for arsenic, and where analogous measurements for other elements are represented similarly.

The pooled, within-bullet standard deviation, *SD(As)*, is computed as follows:

$$SD(As) = \sqrt{\frac{\sum_{i=1}^{200}\sum_{j=1}^{3}(As_i^i - As_i^+)(As_i^j - As_i^+)/2}{200}}$$

(where the 200 in the denominator is for bullets from a single manufacturer). Similarly, the pooled *covariance* between the measurements for two elements, such as arsenic and cadmium, is:

$$Cov(As, Cd) = \frac{\sum_{i=1}^{200}\sum_{j=1}^{3}(As_i^j - As_i^+)(Cd_i^j - Cd_i^+)/2}{200}$$

and similarly for other pairs of elements. The covariance is used to calculate the pooled, within-bullet correlation, defined as follows:

$$Corr(As, Cd) = \frac{Cov(As, Cd)}{SD(As)SD(Cd)}$$

*SD(As)* is more accurate than the within-bullet standard deviations defined for a single bullet above since these estimates are pooled, or averaged, over 200 bullets rather than three fragments. However, the pooling utilizes an assumption of homogeneous variances across bullets, which needs to be justified. (See Appendix F for details.) One aspect of this question was examined by separately computing the within-bullet standard deviations and correlations, as shown above, for each of the four manufacturers. The results of this analysis are also

TABLE 3.2 Pooled Estimates of Within-Bullet Relative Standard Deviations of Concentrations

| | As | Sb | Sn | Bi | Cu | Ag | Cd |
|---|---|---|---|---|---|---|---|
| 800 bullets, %[a] | 5.1 | 2.1 | 3.3 | 4.3 | 2.2 | 4.6 | — |
| 1,837 bullets, 100 × med(SD/avg), % | 10.9 | 1.5 | 118.2 | 2.4 | 2.0 | 2.0 | 33.3 |

[a] Note: All RSDs based on ICP-OES measurements. RSDs for As and Sn based on 200 Federal bullets. RSDs for Sb, Bi, Cu, and As based on within-bullet variances averaged across four manufacturers (800 bullets). Estimated RSD for NAA-As is 5.1 percent.

given in Appendix F. There it is shown that the standard deviations are approximately equal across manufacturers.

The pooled within-bullet standard deviations on the logarithmic scale (or RSDs) for the 800-bullet and 1,837-bullet data sets are given in Table 3.2. Nearly all of the within-bullet standard deviations are between 2 and 5 (that is, between 2 and 5 percent of the mean on the original scale), a range that is narrow enough to consider the possibility that substantially more variable data might have been excluded.

The estimated (pooled) within-bullet correlations, in Table 3.3, are all positive, but many are close to zero, which indicates that for those element pairs, measurements that are high (or low) for one element are generally not predictive of high or low measurements for others. Four notable cases where the correlations are considerable are those between the measurements for Sb and Cu, estimated as 0.67, and the correlations between the measurements for Ag and Sb, Ag and Cu, and Sb and Bi, all estimated as between 0.30 and0.32. Since the full 800-bullet data set provided only five of the seven elements of interest, there are $\binom{5}{2} = 10$ distinct correlations, with the four mentioned above higher than 0.30, two more between 0.10 and 0.30, and four less than 0.10.

TABLE 3.3 Within-Bullet Correlations (800-Bullet Data Set)

| | Average within-bullet correlation matrix | | | | |
|---|---|---|---|---|---|
| | NAA-As | ICP-Sb | ICP-Cu | ICP-Bi | ICP-Ag |
| NAA-As | 1.00 | 0.05 | 0.04 | 0.03 | 0.04 |
| ICP-Sb | 0.05 | 1.00 | 0.67 | 0.32 | 0.31 |
| ICP-Cu | 0.04 | 0.67 | 1.00 | 0.26 | 0.30 |
| ICP-Bi | 0.03 | 0.32 | 0.26 | 1.00 | 0.16 |
| ICP-Ag | 0.04 | 0.31 | 0.30 | 0.16 | 1.00 |

It has been commonly assumed that within-bullet measurements are uncorrelated (or independent), but these data suggest that this assumption is not appropriate. These observed correlations could be due to the measurement process, or possibly different manufacturing processes used by the four suppliers for different lots of lead. Positive correlations, if real, will bias the estimated rate of false matches and false non-matches for statistical procedures that rely on the assumption of zero correlations or independence, and the bias might be substantial. The bias would likely be in the direction of increasing the probability of a false match. That is, error rates calculated under the assumption of independence would tend to be lower than the true rates if there is positive correlation. In particular, probabilities for tests, such as the 2-SD overlap procedure, that operate at the level of individual elements and then examine how many individual tests match or not, cannot be calculated by simply multiplying the individual element probabilities, since the multiplication of probabilities assumes independence of the separate tests.

Since the 1,837-bullet data set used by the committee does not include multiple measurements per bullet (only summary averages and standard deviations), it could not be used to estimate within-bullet correlations. However, the standard deviations of the three measurements that are given provide information on within-bullet standard deviations that can be compared to those from the 800-bullet data set. Medians of the bullet-specific within-bullet standard deviations from the 1,837-bullet data set (actually RSDs) can be compared to those pooled across the 800-bullet data set. The comparisons are given in Table 3.2.[18] While there appears to be fairly strong agreement between the two data sets, there is a severe discrepancy for Sn, which is the result of a small number of outlying values in the 1,837-bullet data set. Again, the existence of outliers is not a property of a normal distribution (outliers are *defined* by not belonging to the assumed distribution), and therefore procedures that are overly reliant on the assumption of normality are potentially misleading.

We have referred to the possible bias of using a subset of the 71,000-bullet data set selected so that it was likely to be more heterogeneous than a full subset of bullets drawn from different melts. This possible bias should be investigated. Further, since the measurement of within-bullet standard deviations and correlations is central to the assessment of operating characteristics of testing procedures, it is unfortunate that the availability of multiple measurements (three measurements on three fragments) on each bullet were not reported in the 1,837-bullet data set. An analysis to verify the estimates of the within-bullet standard deviations and the within-bullet correlations should be carried out if the 71,000 bullet

[18] On occasion, when the three fragment averages were virtually identical, the FBI substituted a minimum measurement based on the instrumentation in place of the RSD.

data are structured in a way that makes this computation straightforward. If the data are not structured in that way, or if the data have not been retained, data for all nine measurements that are collected in the future should be saved in a format that enables these computations to be carried out.

More generally, a philosophical view of this problem is to consider bullet lead heterogeneity occurring to a lesser degree as one gets to more disaggregate bullet lead volumes. Understanding how this decrease occurs would help identify procedures more specific to the problem at hand. Some of this understanding would result from decomposing the variability of bullet lead into its constituent parts, i.e., within-fragment variation (standard deviations and correlations), between-fragment within-bullet variation, between-bullet within-wire reel variation, between-wire reel and within-manufacturer variation, and between-manufacturer variation. Though difficult to do comprehensively, and recognizing that data sets are not currently available to support this, partial analyses that shed light on this decomposition need to be carried out when feasible.

### Between-Bullet Standard Deviations and Correlations

The previous section examined within-bullet standard deviations and correlations, that is, standard deviations and correlations concerning multiple measurements for a single bullet. These statistics are useful in modeling the types of consistency measures that one could anticipate observing from CS and PS bullets from the same CIVL. To understand how much bullets from different CIVLs differ, and the impact on consistency measures, one needs information about the standard deviations and correlations of measurements of bullets from different CIVLs.

The primary source of this information is the 1,837-bullet data set transformed to the logarithmic scale. If the 1,837-bullet data set were a random sample of the population of bullets from different CIVLs, an estimate of the standard deviation across bullets, for, say, arsenic, would be given by:

$$SD^{Across}(As) = \sqrt{\sum_{i=1}^{1837} (As_i^+ - As_+^+)(As_i^+ - As_+^+) / 1,836}$$

and an estimate of the correlation between two elements—say, Ag and Sb—would be given by:

$$\frac{\sum_{i=1}^{1837} (As_i^+ - As_+^+)(Cd_i^+ - Cd_+^+) / 1,836}{(SD^{Across}(As))(SD^{Across}(Cd))}$$

TABLE 3.4 Between-Bullet Standard Deviations (Log Scale) and Correlations (1,837-Bullet Data Set)

| Stand. Devs: | As 4.52 | Sb 4.39 | Sn 5.79 | Bi 1.33 | Cu 2.97 | Ag 1.16 | Cd 2.79 |
|---|---|---|---|---|---|---|---|
| Correlations: | As | Sb | Sn | Bi | Cu | Ag | Cd |
| As | 1.00 | 0.56 | 0.62 | 0.15 | 0.39 | 0.19 | 0.24 |
| Sb | 0.56 | 1.00 | 0.45 | 0.16 | 0.36 | 0.18 | 0.13 |
| Sn | 0.62 | 0.45 | 1.00 | 0.18 | 0.20 | 0.26 | 0.18 |
| Bi | 0.15 | 0.16 | 0.18 | 1.00 | 0.12 | 0.56 | 0.03 |
| Cu | 0.39 | 0.36 | 0.20 | 0.12 | 1.00 | 0.26 | 0.11 |
| Ag | 0.19 | 0.18 | 0.26 | 0.56 | 0.26 | 1.00 | 0.08 |
| Cd | 0.24 | 0.13 | 0.18 | 0.03 | 0.11 | 0.08 | 1.00 |

where, e.g., $As^{+}_{+}$ is the average over fragments and over bullets of the composition of arsenic in the data set (with smaller sample sizes in the case of missing observations). Acknowledging the possible impact of the non-random selection, Table 3.4 provides estimates of the between-bullet standard deviations on the logarithmic scale.

Table 3.4 also displays the between-bullet sample correlation coefficients from the 1,837-bullet data set. All correlations are positive and a few exceed 0.40. In particular, the correlation between Sn and As is .62. Therefore, when one has a bullet that has a high concentration of Sn relative to other bullets, there is a substantial chance that it will also have a high concentration of As.

### Further Discussion of Bullet Homogeneity Using Randich data set

The data in the Randich bullet data set were collected to compare the degree of heterogeneity between and within lead casting, from which bullets are manufactured. Appendix G presents an analysis of those data. Here we focus on comparing the within-measurement standard deviations obtained using the 800-bullet data set with the within-lot standard deviations in the Randich data. The former includes five of the seven elements (As, Sb, Cu, Bi, and Ag), calculated, as before, on the logarithms of the original measurements, and so they are essentially equal to the RSDs on the original scale of measurement. The results are presented in Table 3.5.

For concentrations of the elements As and Sb, the variability of the three measurements from a lot (beginning, middle, and end; or B, M, and E) is about the same as the variability of the three measurements per bullet in the 800-bullet data set. For Bi and Ag, the within-lot variability (B, M, and E) is much smaller than the within-bullet variability in the 800-bullet data set; this finding is unexpected. Further investigation is needed to verify this finding and to determine how and why variation within a bullet could be larger than variation from end to end of a lot from which bullets are made. The within-lot standard deviation of

TABLE 3.5 Comparison of Within-Bullet and Within-Lot Standard Deviations[a]

| | As | Sb | Cu | Bi | Ag |
|---|---|---|---|---|---|
| Between lots: Randich et al. | .706 | .064 | .423 | .078 | .209 |
| Within-bullet: 800-bullet data | .051 | .021 | .022 | .043 | .046 |
| Within-lots: Randich et al. | .056 | .018 | .029 | .008 | .017 |
| Ratio of within-lot to within-bullet: | 1.1 | 0.9 | 1.3 | 0.2 | 0.4 |

[a]Note that the within-lot standard deviation for Cu (column 3) is based on only 23 of the 28 lots, excluding lots 423, 426, 454, 464, 465, which were highly variable. The within-lot standard deviation using all 28 lots is .144.

the three Cu measurements is larger than the within-bullet standard deviation obtained in the 800-bullet data set because of some very unusual measurements in five lots; when these are excluded, the estimated within-lot standard deviation is similar to the within-bullet standard deviation in the 800-bullet data set. Again, further investigation is needed to determine whether this large within-CIVL variance for copper is a general phenomenon, and if so, how it should affect interpretations of bullet lead data. Randich et al. (2002) do not provide replicates or precise within-replicate measurement standard errors, so one cannot determine whether the precision of one of their measurements is equivalent to the precision of one of the FBI measurements.

The above table can also be used to compare lot-to-lot variability to within-lot variability. For four of the five elements, the lot-to-lot variability was 9–15 times greater than within-lot variability. Finally, separate two-way analyses of variance on the logarithms of the measurements on six elements, using the two factors "lot" and "position in lot," show that the position factor for five of the six elements (all but Sn) is not statistically significant at the $\alpha = 0.05$ level. So the variability between lots greatly dominates the variability within lot. The significance for Sn results from two extreme values in this data set, both occurring at the end (namely, $B = M = 414$ and $E = 21$; and $B = 377$, $M = 367$, and $E = 45$). Some lots also yielded three highly dispersed Cu measurements, for example, $B = 81$, $M = 104$, and $E = 103$, and $B = 250$, $M = 263$, and $E = 156$. In general, no consistent patterns (such as, $B < E < M$ or $E < M < B$) are discernible for measurements within lots on any of the elements, and, except for five lots with highly dispersed Cu, the within-lot variability is about the same as or smaller than the measurement uncertainty (see Appendix G for details).

Overall, the committee finds a need for further investigation of the variability of these measurements as a necessary tool for understanding measurement uncertainty and between-CIVL variability, which will affect the assessment of matches between bullets.

**Differences in Average Concentrations—The Relative Mean Difference**

The distribution of concentrations among bullets is important for understanding the differences that need to be identified by the testing procedures, i.e., what differences exist between pairs of unrelated bullets that should result in the pair being excluded from those judged to be matches. We have already examined between-bullet standard deviations and correlations. This section is devoted to the average relative difference in chemical composition of bullets manufactured from different CIVLs. This is related to the between-bullet standard deviations, but is on a scale that is somewhat easier to interpret. There are two sources of information on this: the 1,837-bullet data set and the data in Table 1 of Randich et al. (2002). Both of these sources provide some limited information on differences in average concentrations between bullets from different lead castings (in the case of Randich et al.) or other sources (as suggested by the FBI for the 1,837-bullet data set.) The difference in the average concentration relative to the measurement uncertainty is quite large for most pairs of bullets, but it sometimes happens that bullets from different sources have differences in average concentrations that are within the measurement uncertainty, i.e., the within-bullet or within-wire reel standard deviation.

For example, lots 461 and 466 in Table 1 of Randich et al. (2002) showed average concentrations of five of the six elements roughly within 3–7 percent of each other:

| | Sb | Sn | Cu | As | Bi | Ag |
|---|---|---|---|---|---|---|
| 461 (average) | 696.3 | 673.0 | 51.3 | 199.3 | 97.0 | 33.7 |
| 466 (average) | 721.0 | 632.0 | 65.7 | 207.0 | 100.3 | 34.7 |
| % difference | –3.4% | 6.4% | –21.8% | –3.7% | –3.3% | –2.9% |

These data demonstrate that two lots may differ by as little as a few percent in at least five of the elements currently measured in CABL analysis.

Further evidence that small differences can occur between the average concentrations in two apparently different bullets arises in the closest 47 pairs of bullets among the 854 bullets in the 1,837-bullet data set in which all seven elements were measured (364,231 possible pairs). For 320 of the 329 differences between elemental concentrations (47 bullet pairs, each with 7 elements = 329 element comparisons), the difference is within a factor of 3 of the measurement uncertainty. That is, if the measured difference in mean concentrations (estimated by the difference in the measured averages) is $\delta$ and $\sigma$ = measurement uncertainty (estimated by a pooled within-bullet standard deviation), an estimate of $\delta/\sigma$ is less than or equal to 3 for 320 of the 329 element differences. For three of the bullet pairs, the *relative mean difference* (RMD), the difference in the sample means divided by the larger of the within-bullet standard deviations, is *less than 1 for all seven elements*. For 30 pairs, the RMD is less than or equal to 3, again for all seven elements. So, although the mean concentrations of elements in most of the 854 bullets (selected from the 1,837-bullet data set) often differ by a factor that is many times greater than the measurement uncertainty, some of these unrelated pairs of bullets, selected by the FBI to be from distinct scenarios, show mean differences that can be as small as 1 to 3 times the measurement uncertainty.

## ESTIMATING THE FALSE MATCH PROBABILITIES OF THE FBI'S TESTING PROCEDURES

We utilize the notation developed earlier, where $CS_i$ represented the average of three measurements of the $i^{th}$ fragment of the crime scene bullet, and similarly for $PS_i$. We again assume that there are seven of these sets of measures, corresponding to the seven elements. These measurements are logarithmic transformations of the original data. As before, consider the following statistics:

$$avg(CS) = \frac{CS_1 + CS_2 + CS_3}{3},$$

the overall average over the three pieces for the CS bullet,

$$sd(CS) = \sqrt{\frac{(CS_1 - avg(CS))^2 + (CS_2 - avg(CS))^2 + (CS_3 - avg(CS))^2}{2}},$$

the standard deviation for the CS bullet, and the

$$range(CS) = \max(CS_1, CS_2, CS_3) - \min(CS_1, CS_2, CS_3).$$

The analogous statistics are computed for the PS bullet.

The 2-SD interval for the CS bullet is: $(avg(CS) - 2sd(CS), avg(CS) + 2sd(CS))$, and the 2-SD interval for the PS bullet is: $(avg(PS) - 2sd(PS), avg(PS) + 2sd(PS))$. The range for the CS bullet is: $[\min(CS_1,CS_2,CS_3), \max(CS_1,CS_2, CS_3)]$ and the range for the PS bullet is: $[\min(PS_1,PS_2,PS_3), \max(PS_1,PS_2,PS_3)]$. We denote the unknown true concentration for the CS bullet as $\mu(CS)$, and the unknown true concentration for the PS bullet as $\mu(PS)$. We also denote the unknown true standard deviation for both CS and PS as $\sigma$.[19] Finally, define $\delta = \mu(CS) - \mu(PS)$, the difference between the true concentrations. We do not expect $avg(CS)$ to differ from the true concentration $\mu(CS)$ by much more than twice the standard deviation of the mean $\frac{2\sigma}{\sqrt{3}} \approx 1.15\sigma$, and similarly for PS, though there is a probability of about 10 percent that one or both differ by this much or more.

[19]To estimate the joint measurement uncertainty, we use: $sd = \sqrt{\frac{sd(CS)^2 + sd(PS)^2}{2}}$.

Similarly, we do not expect *avg*(*CS*) – *avg*(*PS*) to differ from the true difference in means δ by much more than $2\sqrt{\sigma^2/3+\sigma^2/3} \approx 1.6\sigma$, though it will happen occasionally.

One of the two errors that can be made in this situation is to falsely judge the CS and PS bullets to be matches when they come from distinct CIVLs. We saw in the previous section that bullets from different CIVLs can have, on occasion, very similar chemical compositions. Since in many cases a match will be incriminating, we would like to make the probability of a false match small.[20] We therefore examine how large this error rate is for both of the FBI's current procedures, and to a lesser extent, for chaining. This error rate for false matches, along with the error rate for false non-matches, will be considerations in suggesting alternative procedures. To start, we discuss the FBI's calculation of the rate of false matching.

### FBI's Calculation of False Match Probability

The FBI reported an estimate of the false match rate through use of the 2-SD-overlap test procedure based on the 1,837-bullet data set. (Recall that this data set has a considerable amount of missing data.) The committee replicated the method on which the FBI's estimate was based as follows. For each of the 1.686 million, i.e., $\binom{1,837}{2}$ pairs of bullets from this data set, the 2-SD overlap test was used to determine whether each pair matched. It was found that 1,393 bullets matched no others, 240 bullets matched one other, 97 bullets matched two others, 40 bullets matched three others, and 12 bullets matched four others. In addition, another 55 bullets matched from 5 to 33 bullets. (The maximum was achieved for a bullet that only had three chemical concentrations measured.) A total of 693 unique pairs of bullets were found to match, which gives a probability of false match of 693/1.686 million = 1/2,433 or .04 percent. As mentioned above, this estimate may be biased low because the 1,837 bullets were selected in part in an attempt to choose bullets from different CIVLs.

It is important to understand the concept of a random sample of bullets in this context. Many different domestic manufacturers make bullets that are used in the United States, and a small proportion of bullets sold in the United States are from foreign manufacturers. Bullets are used in a number of activities, including sport, law enforcement, hunting, and criminal activity, and there may be differences in bullet use by manufacturer. (See Carriquiry et al., 2002, for

[20]We note that bullet lead matching, like DNA matching, may be exonerating. For example, when there are multiple suspects, a match with bullets possessed by one of them would be evidence exonerating the others.

relevant analysis of this point.) While it may make no appreciable difference, it may be useful to consider what the correct reference population of bullets is for this problem. Once that has been established, one could then consider how to sample from that reference population or a closely related population, since it may be the case that sampling would be easier to carry out for a population that was slightly different from the reference population, and deciding to do so might appropriately trade off sampling feasibility for a very slight bias. One possible reference population is all bullets collected by the FBI in criminal investigations. However, a reference population should be carefully chosen, since the false match and non-match rates can depend on the bullet manufacturer and the bullet type. One may at times restrict one's attention to those subpopulations.

### Simulating False Match Probability

The panel carried out a simulation study to estimate the false match rate of the FBI's procedures. Three measurements, normally distributed with mean one and standard deviation $\sigma$ were randomly drawn using a standard pseudo-random number generator to represent the measurements for a CS bullet, and similarly for the PS bullet, except that the mean in the latter case was $1 + \delta$, so that the relative change in the mean is $\delta$. The panel then computed both the 2-SD intervals and the range intervals and examined whether the 2-SD intervals overlapped or the range intervals overlapped, in each case indicating a match. This was independently simulated 100,000 times for various values of $\sigma$ (0.005, 0.010, 0.015, 0.020, 0.025, and 0.030) and various values of $\delta$ (0.0, 0.1, 0.2, …, 7.0). The choices for $\sigma$ were based on the estimated within-bullet standard deviations of less than .03, or 3.0 percent. The choices for $\delta$ were based on the data on differences in average concentrations between bullets. Clearly, except for the situations where $\delta$ equals zero, the (false) match probability should be small. (In Appendix F, it is shown that this probability is a function of only the ratio $\delta/\sigma$. Also, "1" for the mean concentration in the CS bullet is chosen for simplicity and does not reduce the generality of conclusions.)

The sample standard deviation is not unbiased as an estimate of the true standard deviation; its average value (when it is calculated from three normal observations) is $0.8862\sigma$. Therefore, when the sample means of the CS and the PS bullets lie within four times this distance, or $2(sd(CS) + sd(PS))$, which is approximately $2(0.8862\sigma + 0.8862\sigma) = 3.55\sigma$, the 2-SD intervals will overlap. Because the allowance for the difference in sample means is only $1.6\sigma$ given typical error levels for hypothesis testing (see above), the FBI allowance of approximately $3.55\sigma$ being more than twice as wide raises a concern that the resulting false match and false non-match probabilities do not represent a trade-off of these error rates that would be considered desirable. (Note that for the normal distribution, the probability drops off rapidly outside of the range of two standard deviations but not for longer-tailed distributions.) For ranges, under the assumption of normality, a rough computation shows that the ranges will overlap when the sample means lie within 1.69σ of each other, which will result in a lower false match rate than for the 2-SD overlap procedure.

The resulting estimates of the false match rates from this simulation for eight values of δ(0, 1, 2, 3, 4, 5, 6, and 7) and for six values of σ (0.005, 0.01, 0.015, 0.020, 0,025, and 0.030) are shown in Table 3.6 and Table 3.7. Note that the column δ = 0 corresponds to the situation where there is no difference in composition between the two bullets, and is therefore presenting a true match probability, not a false match probability.

For seven elements, the 2-SD-overlap and range-overlap procedures declare a false match only if the 2-SD intervals (or ranges) overlap on all seven elements. If the true difference in all element concentrations were equal (for example, δ = 2.0 percent for all seven elements), the measurement uncertainty were constant for all elements (for example, σ = 1.0 percent), and the measurement errors for all seven elements were independent, the false match probability for seven elements would equal the product of the per-element rate seven times (for example, for δ = 2.0, σ = 1.0, $.841^7 = 0.298$ for the 2-SD-overlap procedure, and $.377^7 = 0.001$ for the range-overlap procedure). Tables 3.8 and 3.9 give the corresponding false match probabilities for seven elements, assuming independence among the measurement errors on all seven elements.

The false match probabilities in Tables 3.8 and 3.9 are lower bounds because the analysis in the previous section indicated that the measurement errors are likely not independent. Thus, the actual seven-element false match probabil-

TABLE 3.6 False Match Probabilities with 2-SD-Overlap Procedure, One Element (δ = 0–7%, σ = 0.5–3.0%)

| σ/δ | 0 | 1 | 2 | 3 | 4 | 5 | 6 | 7 |
|---|---|---|---|---|---|---|---|---|
| 0.5 | 0.990 | 0.841 | 0.369 | 0.063 | 0.004 | 0.000 | 0.000 | 0.000 |
| 1.0 | 0.990 | 0.960 | 0.841 | 0.622 | 0.369 | 0.172 | 0.063 | 0.018 |
| 1.5 | 0.990 | 0.977 | 0.932 | 0.841 | 0.703 | 0.537 | 0.369 | 0.229 |
| 2.0 | 0.990 | 0.983 | 0.960 | 0.914 | 0.841 | 0.742 | 0.622 | 0.495 |
| 2.5 | 0.990 | 0.986 | 0.971 | 0.944 | 0.902 | 0.841 | 0.764 | 0.671 |
| 3.0 | 0.990 | 0.987 | 0.978 | 0.960 | 0.932 | 0.892 | 0.841 | 0.778 |

TABLE 3.7 False Match Probabilities with Range-Overlap Procedure, One Element (δ = 0–7%, σ = 0.5–3.0%)

| σ/δ | 0 | 1 | 2 | 3 | 4 | 5 | 6 | 7 |
|---|---|---|---|---|---|---|---|---|
| 0.5 | 0.900 | 0.377 | 0.018 | 0.000 | 0.000 | 0.000 | 0.000 | 0.000 |
| 1.0 | 0.900 | 0.735 | 0.377 | 0.110 | 0.018 | 0.002 | 0.000 | 0.000 |
| 1.5 | 0.900 | 0.825 | 0.626 | 0.377 | 0.178 | 0.064 | 0.018 | 0.004 |
| 2.0 | 0.900 | 0.857 | 0.735 | 0.562 | 0.377 | 0.220 | 0.110 | 0.048 |
| 2.5 | 0.900 | 0.872 | 0.792 | 0.672 | 0.524 | 0.377 | 0.246 | 0.148 |
| 3.0 | 0.900 | 0.882 | 0.825 | 0.735 | 0.626 | 0.499 | 0.377 | 0.265 |

TABLE 3.8 False Match Probabilities with 2-SD-Overlap Procedure, Seven Elements (Assuming Independence: $\delta$ = 0–7%, $\sigma$ = 0.5–3.0%)

| $\sigma/\delta$ | 0 | 1 | 2 | 3 | 4 | 5 | 6 | 7 |
|---|---|---|---|---|---|---|---|---|
| 0.5 | 0.931 | 0.298 | 0.001 | 0.000 | 0.000 | 0.000 | 0.000 | 0.000 |
| 1.0 | 0.931 | 0.749 | 0.298 | 0.036 | 0.001 | 0.000 | 0.000 | 0.000 |
| 1.5 | 0.931 | 0.849 | 0.612 | 0.303 | 0.084 | 0.013 | 0.001 | 0.000 |
| 2.0 | 0.931 | 0.883 | 0.747 | 0.535 | 0.302 | 0.125 | 0.036 | 0.007 |
| 2.5 | 0.931 | 0.903 | 0.817 | 0.669 | 0.487 | 0.302 | 0.151 | 0.062 |
| 3.0 | 0.931 | 0.911 | 0.850 | 0.748 | 0.615 | 0.450 | 0.298 | 0.175 |

TABLE 3.9 False Match Probabilities with Range-Overlap Procedure, Seven Elements (Assuming Independence: $\delta$ = 0–7%, $\sigma$ = 0.5–3.0%)

| $\sigma/\delta$ | 0 | 1 | 2 | 3 | 4 | 5 | 6 | 7 |
|---|---|---|---|---|---|---|---|---|
| 0.5 | 0.478 | 0.001 | 0.000 | 0.000 | 0.000 | 0.000 | 0.000 | 0.000 |
| 1.0 | 0.478 | 0.116 | 0.001 | 0.000 | 0.000 | 0.000 | 0.000 | 0.000 |
| 1.5 | 0.478 | 0.258 | 0.037 | 0.001 | 0.000 | 0.000 | 0.000 | 0.000 |
| 2.0 | 0.478 | 0.340 | 0.116 | 0.018 | 0.001 | 0.000 | 0.000 | 0.000 |
| 2.5 | 0.478 | 0.383 | 0.197 | 0.062 | 0.011 | 0.001 | 0.000 | 0.000 |
| 3.0 | 0.478 | 0.415 | 0.261 | 0.116 | 0.037 | 0.008 | 0.001 | 0.000 |

ity is likely to be higher than the false match probabilities for a single element raised to the seventh power, which are what are displayed. As shown below, the panel has determined that for most cases the correct false match probability will be closer to the one element probability raised to the fifth or sixth power.

Table 3.8 for the 2-SD-overlap procedure for seven elements is rather disturbing in that for values of $\delta$ around 3.0, indicating fairly sizeable differences in concentrations, and for reasonable values of $\sigma$, the false match probabilities can be quite substantial. (A subset of the 1,837-bullet data set showed only a few pairs of bullets where $\delta/\sigma$ might be as small as 3 for all seven elements. However, the 1837-bullet data set was constructed to contain bullets selected to be as distinct as possible, so the actual frequency is likely higher.)

A simulation study using the within-bullet correlations from the Federal bullets and assuming the Cd measurement is uncorrelated with the other six elements suggests that the false match probability is close to the single element rate raised to the fifth power. An additional simulation study carried out by the panel, based on actual data, further demonstrated that the false match probabilities on seven elements are likely to be higher than the values shown in Table 3.8 and 3.9. The study was conducted as follows:

1. Select a bullet at random from among the 854 bullets (of the 1,837 bullet data set) in which all seven elements were measured.

TABLE 3.10 Simulated False Match Probabilities Based on Real Data[a]

| | δ | | | |
|---|---|---|---|---|
| Method | 3% | 5% | 7% | 10% |
| 2-SD overlap | 0.404 | 0.273 | 0.190 | 0.127 |
| Range overlap | 0.158 | 0.108 | 0.053 | 0.032 |

[a]Note that the columns represent differences in bullets that are relatively small given the distribution of between-bullet differences from the 1,837-bullet data set. One would expect the false match probability to be smaller for larger differences between bullets.

2. Start with seven independent standard normal variates. Transform these seven numbers so that they have the same correlations as the estimated within-bullet correlations. Multiply the individual transformed values by the within-bullet standard deviations to produce a multivariate normal vector of bullet lead concentrations with the same covariance structure as estimated using the 200 Federal bullets in the 800-bullet data set. Add these values to the values for the randomly selected bullet. Repeat this three times to produce the three observations for the CS bullet. Repeat this for the PS bullet, except add δ to the values at the end.

3. For each bullet calculate the within-bullet means and standard deviations, and carry out the 2-SD-overlap and range-overlap procedures.

4. Repeat 100,000 times, calculating the overall false match probabilities for four values of δ, 0.03, 0.05, 0.07, and 0.10.

The results of this simulation are given in Table 3.10.

Generally speaking, the false match probabilities from this simulation were somewhat higher than those given in Tables 3.8 and 3.9. This may be due to either a larger than anticipated measurement error in the 854 bullet data set, the correlations among the measurement errors, or both. (This simulation does not include false matches arising from the possibility of two CIVLs having the same composition.)

This discussion has focused on situations in which the means for the CS and PS bullets were constant across elements. For the more general case, the results are more complicated, though the above methods could be used in those situations.

### False Match Probability for Chaining

To examine the false match probability for chaining, the panel carried out a limited analysis. The FBI, in its description of chaining, states that one should avoid having a situation in which bullets in the reference population form compositional groups that contain large numbers of bullets. (It is not clear how the algorithm should be adjusted to prevent this from happening.) This is because large groups will tend to have a number of bullets that as pairs may have concentrations that are substantially different.

To see the effect of chaining, consider bullet 1,044, selected at random from the 1,837-bullet data set. The data for these bullets are given in the first two lines of Table 3.11.

Bullet 1,044 matched 12 other bullets; that is, the 2-SD interval overlapped on all elements with the 2-SD interval for 12 other bullets. In addition, each of the 12 other bullets in turn matched other bullets; in total, 42 unique bullets were identified. The variability in the averages and the standard deviations of the 42 bullets would call into question the reasonableness of placing them all in the same compositional group. The overall average and average standard deviation of the 42 average concentrations of the 42 "matching" bullets are given in the third and fourth lines of Table 3.11. In all cases, the average standard deviations are at least as large as, and usually 3–5 times larger than, the standard deviation of bullet 1,044, and larger standard deviations are associated with wider intervals and hence more false matches. Although this illustration does not present a comprehensive analysis of the false match probability for chaining, it demonstrates that this method of assessing matches could possibly create more false matches than either the 2-SD-overlap or the range-overlap procedures.

One of the questions presented to the committee (see Chapter 1) was, "Can known variations in compositions introduced in manufacturing processes be used to model specimen groupings and provide improved comparison criteria?" Bullets from the major manufacturers at a specific point in time might be able to be partitioned based on the elemental compositions of bullets produced. However, there are variations in the manufacturing process by hour and by day, there are a large number of smaller manufacturers, and there may be broader trends in composition over time. These three factors will erode the boundaries between these partitions. Given this and the reasons outlined above, chaining is unlikely to serve the desired purposes of identifying matching bullets with any degree of reliability. In part due to the many diverse methods that could be applied, the panel has not examined other algorithms for partitioning or clustering bullets to determine whether they might overcome the deficiencies of chaining. FBI support for such a study may provide useful information and a more appropriate partitioning algorithm that has a lower false match rate than chaining appears to have.

TABLE 3.11 Elemental Concentrations for Bullet 1,044

| | As | Sb | Sn | Bi | Cu | Ag | Cd |
|---|---|---|---|---|---|---|---|
| Average | 0.0000 | 0.0000 | 0.0000 | 0.0121 | 0.00199 | 0.00207 | 0.00000 |
| SD | 0.0002 | 0.0002 | 0.0002 | 0.0002 | 0.00131 | 0.00003 | 0.00001 |
| Avg of 42 Avgs | 0.0004 | 0.0004 | 0.0005 | 0.0110 | 0.00215 | 0.00208 | 0.00001 |
| SD of 42 Avgs | 0.0006 | 0.0005 | 0.0009 | 0.0014 | 0.00411 | 0.00017 | 0.00001 |

### Alternative Testing Strategies

We have discussed the strategies used by the FBI to assess match status. An important issue is the substantial false match rate that occurs when using the 2-SD overlap procedure for bullets with elemental compositions that differ by amounts moderately larger than the within-bullet standard deviation. (This concern arises to a somewhat lesser degree for the range overlap procedure.) In addition, all three of the FBI's procedures fail to represent current statistical practice, and as a result the data are not used as efficiently as they would be if the FBI were to adopt one of the alternative test strategies proposed for use here. A result of this inefficiency is either false match rates, false non-match rates, or both, that are larger than they could otherwise be.

This section describes alternative approaches to assessing the match status for two bullets, CS and PS, in a manner that makes effective and efficient use of the data collected, so that neither the false match nor the false non-match rates can be made smaller without an increase in the other, and so that estimates of these error rates can be calculated and the reliability of the assessment of match status can be determined.

The basic problem is to judge whether 21 numbers (each an average of three measurements), measuring seven elemental concentrations for each of three bullet fragments from the CS bullet, are either far enough away from the analogous 21 numbers from the PS bullet to be consistent with the hypothesis that the mean concentrations of the CIVLs from which the bullets came are different, or whether they are too close together, and hence more consistent with the hypothesis that the CIVLs means are the same. There are also other data available with information about the standard deviations and correlations of these measurements, and the use of this information is an important issue.

Let us consider one element to start. Again, we denote the three measurements on the CS and PS bullets $CS_1, CS_2, CS_3$ and $PS_1, PS_2, PS_3$, respectively. The basic question is whether three measurements of the concentrations of one of the seven elements from two bullets are sufficently different to be consistent with the following hypothesis, or are sufficiently close to be inconsistent with that hypothesis: that the mean values for the elemental concentrations for the bullets manufactured from the same CIVL with given elemental concentrations, of which the PS bullet is a member, are different from the mean values for the elemental concentrations for the bullets manufactured from a different CIVL of which the CS bullet is a member.

Assuming that the measurements of any one element come from a distribution that is well-behaved (in the sense that wildly discrepant observations are extremely unlikely to occur), and assuming that the standard deviation of the CS measurements is the same as the standard deviation of the PS measurements, the standard statistic used to measure the closeness of the six numbers for this single element is the two sample t-test: $t = \frac{|avg(CS) - avg(PS)|}{\sqrt{[sd(CS)^2 + sd(PS)^2]/3}}$. When additional data on the within-bullet standard deviation is available, whose use we strongly recommend here, the denominator is replaced with a pooled estimate of the assumed common standard deviation $s_p$, resulting in the t-statistic $t = \frac{|avg(CS) - avg(PS)|}{\sqrt{2/3 s_p}}$. To use *t*, one sets a critical value $t_\alpha$ so that when *t* is smaller than $t_\alpha$ the averages are considered so close that the hypothesis of a "non-match" must be rejected, and the bullets are judged to match, and when *t* is larger than $t_\alpha$ the averages are considered to be so far apart that the bullets are judged to not match.

Setting a critical value simultaneously determines a power function. For any given difference in the true mean concentrations for the CS and the PS bullets, δ, there is an associated probability of being judged a match and a probability of being judged a non-match. If δ equals 0, the probability of having *t* exceed the critical value $t_\alpha$ is the probability of a false non-match. If δ is larger than 0, the probability of having *t* smaller than the critical value $t_\alpha$ is the probability of a false match (as a function of δ).

As mentioned early in this chapter, one may also set two critical values to define three regions; match, no decision, and no match. Doing this may have important advantages in helping to achieve error rates that are more acceptable, at the expense of having situations for which no judgment on matching is made. When the assumptions given above obtain (assuming use of the logarithmic transformation), the two-sample *t*-test has several useful properties, given normal data, for study of a single element, and is clearly the procedure of choice. In practice we can check to see how close to normality we believe the bullet data or transformed bullet data are, and if they appear to be close to normality with no outliers we can have confidence that our procedure will behave reasonably.

The spirit of the 2-SD overlap procedure is similar to the two-sample t-test for one element, but results in an effectively much larger critical value than would ordinarily be used because the "SD" is the sum of two standard deviations (SD(CS) + SD(PS)), rather than $s_p\sqrt{2/3}$, which substantially overestimates the standard deviation of the difference between the two sample means. This reduces the false non-match rate when the bullets are identical, and simultaneously increases false match rates when they are different.

To apply the two-sample *t*-test, the only remaining questions are: (a) how to choose $t_\alpha$, and (b) how to estimate the common standard deviation of the measurement error. To estimate the common standard deviation using pooling, it would be necessary to carry out analysis of reference bullets to determine what factors were associated with heterogeneity in within-bullet standard deviations.

Having done that, all reference bullets that could be safely assumed to have equal within-bullet standard deviations could be pooled using the following formula:

$$s_p = \sqrt{\frac{(N_1 - 1)SD_i^2 + \ldots + (N_k - 1)SD_K^2}{N_1 + N_2 + \ldots + N_K - K}}$$

where $N_i$ is the number of replications for the $i^{th}$ bullet used in the computation (typically 3 here), and $K$ is the total number of bullets used for pooling. When $N_i$ is the same for all bullets, (in this application likely $N_1 = N_2 = \ldots = N_K = 3$, then $s_p$ is just the square root of the mean of the squared deviations.

Assuming that the measurements (after transforming using logarithms) are roughly normally distributed, tables exist that, given $t_\alpha$ and $\delta$, provide the false match and false non-match rates. (These are tables of the central and non-central *t* distribution.) Under the assumption of normality, the two-sample *t*-test has operating characteristics—the error rates corresponding to different values of $\delta$—that are as small as possible. That is, given a specific $\delta$, one cannot find a test statistic that has a simultaneously lower false match rate, given a specific $\delta$, and lower false non-match rate.

The setting of $t_\alpha$, which determines both error rates, is not a matter to be decided here, since it is not a statistical question. One can make the argument that the false match rate should be controlled to a level at which society is comfortable. In that case, one would take a particular value of $\delta$, the difference between the CS and PS bullet concentrations that one finds important to discriminate between, and determine $t_\alpha$ to obtain that false match rate, at the same time accepting the associated false non-match rate. Appropriate values of $\delta$ to use will depend on the situation, the manufacturer, and the type of bullet. Having an acceptable false match rate for values where the within-bullet standard deviation becomes unlikely to be a reasonably full explanation for a difference in means would be very beneficial. However, in this case it would still be essential to compute and communicate the false non-match rate, since greatly reducing the false match rate by making $t_\alpha$ extremely small may result in an undesirable trade-off of one error rate versus the other.[21] Further, if one cannot make both error rates as small as would be acceptable, then there may be non-standard steps that can be taken to decrease both error rates, such as taking more readings per bullet or decreasing the measurement error in the laboratory analysis. (This assumes that the main part of within-bullet variability is due to measurement error and not due to within-bullet heterogeneity, which has yet to be confirmed.)

---

[21]It is unlikely for there to be testimony in cases in which there is a non-match, since the evidence will not be included in the case. However, determining this error rate would nevertheless still be valuable to carry out.

Now we add the complication that seven elements are used to make the judgment of match status. The 2-SD overlap procedure uses a unanimous vote for matching based on the seven individual assessments by element of match or non-match status. A problem is that several of the differences for the seven elements may each be close to, but smaller than, the 2-SD overlap criterion, yet in some collective sense, the differences are too large to be compatible with a match. The 2-SD overlap procedure provides no opportunity to accumulate these differences to reach a conclusion of "no match."

To address this, assume first that the within-bullet correlations between elemental concentrations are all equal to zero. In that case, the theoretically optimal procedure, assuming multivariate normality, is to add the squares of the separate *t*-statistics for the seven elements and to use the sum as the test statistic. The distribution of this test statistic is well-known, and false match rates and false non-match rates can be determined for a range of possible critical values and vectors of separation, $\delta$. (There is a separation vector, since with seven elements, to determine a false match rate, one must specify the true distances between the means for the bullets for each of the seven elements.) Again, under the assumptions given, this procedure is theoretically optimal statistically in the sense that no test statistic can have a simultaneously lower false non-match rate and lower false match rate, given a specific separation vector.

However, as seen from the 800-bullet data set, it is apparently not the case that the within-bullet measurements of elemental composition are uncorrelated. If the standard deviations and correlations could be well estimated, the theoretically optimal procedure, assuming multivariate normality, to judge the closeness of the 21 numbers from the CS and the PS bullets would be to use Hotelling's $T^2$ statistic. However, there are three complications regarding the use of $T^2$. First, the within-bullet correlations and standard deviations have not, to date, been estimated accurately. Second, the $T^2$ statistic has best power against alternative hypotheses for which all of the mean elemental concentrations are different between the CS and the PS bullets. If this is not the case, $T^2$ averages the impact of the differences that exist over seven anticipated differences, thus reducing their impact. Given situations where only three or four of the elements exhibit differences, $T^2$ will have a relatively high false match error rate relative to procedures, like 2-SD overlap, that can key on one or two large differences. Third, $T^2$ is somewhat sensitive to large deviations from normality, and the bullet lead data do seem to have frequent outlying observations, whether from heterogeneity within bullets or inadequately controlled measurement methods.

Even given these concerns, once the needed within-bullet correlations have been well-estimated, and the non-(log) normality has been addressed, the use of $T^2$ should be preferred to the use of either the 2-SD overlap or the range overlap procedures. This is because $T^2$ retains the theoretical optimality properties of the simpler tests described above. (It is the direct analogue of the two-sample *t*-test in more than one dimension.) One way to describe the theoretical optimality,

**Theoretical Optimality of $T^2$ Procedure**

Hotelling's $T^2$ uses the observations to calculate the following statistic: $\mathbf{T^2} = n(\mathbf{X} - \mathbf{Y})'S^{-1}(\mathbf{X} - \mathbf{Y})$, without which the $n$ is known as the Mahalanobis distance. This statistic, whether it is used in a formal test or not, has a theoretical optimality property. The same distance between the center (mean) and contours appears in the mathematical formulation of the multivariate normal distribution (in the exponent). This statistic defines "contours" of equal probability around the center of the distribution, and the contours are at lower and lower levels of probability as the statistic increases. This means that, if the observations are multivariate normal, as seems to be approximately the case for the logged concentrations in bullet lead, the probability is most highly concentrated within such a contour. No other function of the data can have this property. The practical result is that the $T^2$ statistic and the chosen value of $T^2_\alpha$ define a region around the observed values of the differences between the PS and the CS bullets that is as small as possible, so that the probability of falsely declaring a match is also as small possible (given a fixed rate for the probability of false non-matches). This is a powerful argument in favor of using the $T^2$ statistic.

given data that are multivariate normal, of $T^2$ is that, for different critical values, say $T^2_\alpha$, $T^2$ defines a region of observed separation vectors that are the most probable if there were no difference between the means of the concentrations of the CS and the PS bullets.

The panel has identified an alternative to the use of the $T^2$ test statistic that retains some of the benefits of being derived from the univariate $t$-test statistic, but also has the advantage of being able to reject a match based on one moderately substantial difference in one dimension, which is an advantage of the 2-SD overlap procedure. This approach, which we will denote the "successive $t$-test approach" test statistics, is as follows:

1. estimate the within-bullet standard deviations for each element using a pooled within-bullet standard deviation $s_p$ from a large number of bullets, as shown above.
2. calculate the difference between the means of the (log-transformed) measurements of the CS and the PS bullets,
3. If all the differences are less than $k_\alpha s_p$ for each of the seven elements for some constant $k_\alpha$, then the bullets are deemed a match, otherwise they are a non-match.

Unfortunately, the estimation of false match rates and false non-match rates for the successive $t$-test statistic is complicated by the lack of independence of within-bullet measurements between the different elements. The panel carried out a number of simulations that support estimating the false match rate by raising the probability of a match for a single element to the fifth power (rather than the seventh power, which would be correct if the within-bullet measurements were independent). That is, raising the individual probabilities to the fifth power provides a reasonable approximation to the true error rates. This is somewhat ad hoc, but further analysis may show that for the modest within-bullet correlations in use, this is a reasonable approximation.[22] In any event, for a specific separation vector, simulation studies can always be used to assess, to some degree of approximation, the false match and false non-match probabilities of this procedure. The advantage of the successive *t*-test statistics are that the approach has the ability to notice single large differences, but also retains the use of efficient measures of variability.

Similar to the above, choices of $k_\alpha$ form a parametric family of test procedures, each of which trades off one of the two error rates against the other. The choice of $k_\alpha$ is again a policy matter that we will not discuss except to stress that whatever the choice of $k_\alpha$ is, if the FBI adopts this suggested procedure, both the false match and the false non-match probabilities must be estimated and communicated in conjunction with the use of this evidence in court.

In summary, the two alternatives to the FBI's test statistics advocated by the panel are the $T^2$ test statistic and the successive *t*-test statistics procedure. If the underlying data are approximately (log) normally distributed, and if pooled estimates, over an appropriate reference set of bullets, are available to estimate within-bullet standard deviations and within-bullet correlations, and finally, if all seven elements are relatively active in discriminating between the CS and the PS bullets, then $T^2$ is an excellent statistic for assessing match status. The successive *t*-test statistics procedure is somewhat less dependent on normality and can be used in situations in which a relatively small number of elements are active. However, quick assessment of error rates involves an approximation. Given the different strengths of these two procedures, there are good reasons to report both results. In addition, the FBI should examine the 71,000, bullet data set for recent data to see whether all seven elements now in use are routinely active, or whether there may be advantages from reducing the elements considered. This would be an extension of the panel's work described above on the 1,837-bullet data set.

In the meantime, both of the recommended approaches have advantages over the use of the current FBI procedures. They are both based on more efficient univariate statistical tests, and they both allow direct estimation (in one case, approximate estimation) of the false match and false non-match rates. One

[22]The FBI should remain open to the possibility, if the within-bullet correlations are higher than current estimates, of dropping one of element pairs involved in very substantial correlations (over .9) to reduce the size of this problem, and to also consider the possibility of adding other elements if differences in those concentrations by manufacturer appear.

procedure, successive *t*-test statistics, is better at identifying non-matching situations in which there are a few larger discrepancies for a subset of the seven elements, and the other, $T^2$, is better at identifying non-matching situations in which there are modest differences for all seven elements. In addition, if $T^2$ is to be used, given the small amount of data collected on the PS and the CS bullets, pooling across a reference data set of bullets to estimate the within-bullet standard deviations and correlations is vital to support this approach.[23]

If both of these procedures are adopted, the FBI must guard against the temptation to compute both statistics and report only the one showing the more favorable result.

We have stressed in several places that prior to use of these test procedures, the operating characteristics, i.e., the false match rate and false non-match rates, be calculated and communicated along with the results of the specific match. (Even though non-matches are unlikely to be presented as evidence in court, knowing the false non-match error rate protects against setting critical values that too strongly favor one error rate against the other.) A different false match rate is associated with each non-zero separation vector $\delta$ (in seven dimensions). It is difficult to prescribe a specific set of separation vectors to use for this communication purpose. However, as in the univariate case, having an acceptable false match rate for separation vectors where the within-bullet standard deviations become unlikely to be a reasonably full explanation for differences in means would be very beneficial. It would also be useful to include a separation vector that demonstrated the performance of the procedure when not all mean concentrations for elements differ.

In addition, for any procedure that the FBI adopts, a much more comprehensive study of the procedure's false non-match and false match rates should be carried out than can be summarized in a small number of false match rates.

In discussing the calculation of false match rates, the panel is devoting its attention to cases that are at least somewhat unclear, since those are the cases for which the choice of procedure is most important. However, for a large majority of bullet pairs that are clearly dissimilar, there would be strong agreement between the procedures that the FBI is using today and the two procedures recommended here as preferred alternatives.

Finally, the 2-SD and range overlap procedures, the $T^2$ test statistic, and to a lesser extent, the successive *t*-test statistics procedure, are all sensitive to the assumption of normality. By sensitive, we mean that the error rates computed under the assumption of (log) normality may be unrealistic if the assumption

---

[23]There is a technical point here, that in using pooled standard deviations and correlations to form the estimated covariance matrix for use with the $T^2$ test statistic, it is important to check that the resulting estimated covariance matrix is positive definite. This is unlikely to be a problem, in this application.

does not hold. Specifically, the presence of outlying values is likely to inflate the estimates of variability more than the differences in concentrations, so that more widely disparate bullet pairs will be found to match using these test statistics. (See Eaton and Efron, 1970; Holloway and Dunn, 1967; Chase and Bulgren, 1971; and Everit, 1979, for the non-robustness of $T^2$.) The FBI could take two actions to address this sensitivity. First, if the non-normality is not a function of laboratory error or contamination or other sources that can be reduced over time, the FBI should use, in addition to the two procedures recommended here, a "robust" test procedure such as a permutation test, to see if there is agreement with the normal-theory based procedure. If there is agreement between the robust and non-robust procedures, one may safely report the results from the standard procedure. If, on the other hand, there is disagreement, the source of the disagreement would need to be investigated to see if outliers or other data problems were at fault. If the non-normality may be a function of human error, the data should be examined prior to use to identify any discrepant measurements so that they can be repeated in order to replace the outlying observation. Identifying outliers from a sample of size three is not easy, but over time, procedures (such as control charts) could be identified that would be effective at determining when additional measurements would be valuable to take.

## RECOMMENDATIONS

The largest source of error in the use of CABL is the unknown variability within the population of bullets in the United States due to variations within and across manufacturing processes. (The manufacturing process and its effect on the interpretation of CABL evidence is discussed in detail in Chapter 4.) This variability is not sufficiently taken into account by the statistical methods currently in use in the analysis of CABL data. In addition, the FBI's methods are not representative of current statistical practice. Several steps can be taken to remedy these problems. A key need is the identification of statistical tests that have acceptable levels of rates of false matches and false non-matches. The committee has proposed a variety of analyses to increase understanding of the variability in the composition of bullet lead, and how to make better use of statistical methods in analyzing this information.

The discussion above supports the following recommendations.

**Recommendation:** The committee recommends that the FBI estimate within-bullet standard deviations on separate elements and correlations for element pairs, when used for comparisons among bullets, through use of pooling over bullets that have been analyzed with the same ICP-OES measurement technique. The use of pooled within-bullet standard deviations and correlations is strongly preferable to the use of within-bullet standard deviations that are calculated from the two bullets being compared. Further, estimated standard deviations should be charted regularly to ensure the stability of the measurement process; only standard deviations within control-chart limits are eligible for use in pooled estimates.

In choosing a statistical test to apply when determining a "match," the goal was to choose a test that had good performance properties as measured by (1) its rate of false non-matches and (2) its rates of false matches, evaluated at a variety of separations between the concentrations of the CS and the PS bullets. The latter corresponds to the probability of providing false evidence of guilt, which our society views as important to keep extremely low.

Given arguments of statistical efficiency that translate into lower error rates, it is attractive to consider either the $T^2$ test statistic, or the successive $t$-test statistics procedure, since they are more representative of current statistical practice. The application of both procedures is illustrated using some sample data in Appendix K.

**Recommendation:** The committee recommends that the FBI use either the $T^2$ test statistic or the successive $t$-test statistics procedure in place of the 2-SD overlap, range overlap, and chaining procedures. The tests should use pooled standard deviations and correlations, which can be calculated from the relevant bullets that have been analyzed by the FBI Laboratory. Changes in the analytical method (protocol, instrumentation, and technique) will be reflected in the standard deviations and correlations, so it is important to monitor these statistics for trends and, if necessary, to recalculate the pooled statistics.

The committee recognizes that some work remains in order to provide additional rigor for the use of this testing methodology in criminal cases. Further exploration of the several issues raised in this chapter should be carried out. As part of this effort, it will be necessary to further mine the extant data resources on lead bullet composition to establish an empirical base for the methodology's use. In addition, this analysis may discover deficiencies in the extant data resources, thereby identifying additional data collection that is needed.

**Recommendation:** To confirm the accuracy of the values used to assess the measurement uncertainty (within-bullet standard deviation) in each element, the committee recommends that a detailed statistical investigation using the FBI's historical data set of over 71,000 bullets be conducted. To confirm the relative accuracy of the committee's recommended approaches to those used by the FBI, the cases that match using the committee's recommended approaches should be compared with those obtained with the FBI approaches, and causes of discrepancies between the two approaches—such as excessively wide intervals from larger-than-expected estimates of the standard deviation, data from specific time periods, or examiners—should be identified. As the FBI adds new bullet data to its

71,000+ data set, it should note matches for future review in the data set, and the statistical procedures used to assess match status.

No matter which statistical test is utilized by examiners, it is imperative that the same statistical protocol be applied in all investigations to provide a replicable procedure that can be evaluated.

**Recommendation:** The FBI's statistical protocol should be properly documented and followed by *all* examiners in *every* case.

## REFERENCES

Carriquiry, A.; Daniels, M.; and Stern, H. "Statistical Treatment of Case Evidence: Analysis of Bullet Lead," *Unpublished report,* Dept. of Statistics, Iowa State University, 2002.

Chase, G.R., and Bulgren, W.G., "A Monte Carlo Investigation of the Robustness of $T^2$," Journal of the American Statistical Association, 1971, 66, pp 499–502.

Eaton, M.L. and Efron, B., "Hotelling's $T^2$ Test Under Symmetry Conditions," Journal of the American Statistical Association, 1970, 65, pp. 702–711.

Everitt, B.S., "A Monte Carlo Investigation of the Robustness of Hotelling's One- and Two-Sample $T^2$ Tests," Journal of the American Statistical Association, 1979, 74, pp 48–51.

Holloway, L.N. and Dunn, O.L., "The Robustness of Hotelling's $T^2$," American Statistical Association Journal, 1967, pp 124–136.

Owen, D.B. "Noncentral t distribution" in *Encyclopedia of Statistical Sciences, Volume 6*; Kotz, S.; Johnson, N. L.; and Read, C.B.; Eds.; Wiley: New York, NY 1985, pp 286–290.

Peele, E. R.; Havekost, D. G.; Peters, C. A.; Riley, J. P.; Halberstam, R. C.; and Koons, R. D. USDOJ, (ISBN 0-932115-12-8), 1991, 57.

Peters, C. A. *Comparative Elemental Analysis of Firearms Projectile Lead by ICP-OES*, FBI Laboratory Chemistry Unit. Issue date: Oct. 11, 2002. *Unpublished* (2002).

Peters, C. A. *Foren. Sci. Comm.* 2002, *4(3)*. <http://www.fbi.gov/hq/lab/fsc/backissu/july2002/peters.htm> as of Aug. 8, 2003.

Randich, E.; Duerfeldt, W.; McLendon, W.; and Tobin, W. *Foren. Sci. Int.* 2002, 127, 174–191.

Rao, C.R., *Linear Statistical Inference and Its Applications*: Wiley, New York, NY 1973.

Tiku, M. "Noncentral F distribution" in *Encyclopedia of Statistical Sciences, Volume 6*; Kotz, S.; Johnson, N. L.; and Read, C.B.; Eds.; Wiley: New York, NY 1985, pp 280–284.

Vardeman, S. B. and Jobe, J. M. *Statistical Quality Assurance Methods for Engineers*, Wiley: New York, NY 1999.

Wellek, S. *Testing Statistical Hypotheses of Equivalence;* Chapman and Hall: New York, NY 2003.

# 4

# Interpretation

The primary objective of compositional analysis of bullet lead (CABL) is to produce evidence for use in court. Although the evidence is analyzed with scientific instrumentation and statistical methods, its presentation and use in court are subject to human interpretation and error. Attorneys, judges, juries, and even expert witnesses can easily and inadvertently misunderstand and misrepresent the analysis of the evidence and its importance. It is therefore essential to discuss whether and how the evidence can be used. It is first necessary to introduce the lead and bullet manufacturing processes so that the implications of bullet production for the legal system are fully understood. This chapter is split into two sections: "Significance of the Bullet Manufacturing Process" and "Compositional Analysis of Bullet Lead as Evidence in the Legal System."

## SIGNIFICANCE OF THE BULLET MANUFACTURING PROCESS

The following description of the processes leading to the production of loaded ammunition represents the bullet manufacturing practices currently in place at large-scale producers in the United States. (Processes used overseas are less well documented.) As shown in this chapter, the processes vary at numerous points, depending on such factors as the manufacturer, the caliber and style of bullet, the magnitude of a production run (which is often dictated by the demand for a particular caliber), and the size of the manufacturing facility. This section details procedures that are believed to account for the manufacturing processes used for .22 caliber rimfire and other bullets by major producers in the United States. (This process is described because .22 caliber rimfire ammunition is one of the most popular ammunition rounds produced.) It has been estimated that 50–75 percent of all ammunition sold in the United States originates with U.S. manufacturers and that about 50 percent of ammunition used by the U.S. military (for example, 9-mm, 7.62-NATO, and 5.56-NATO ammunition) and more than 50 percent of non-U.S. issue military calibers (such as 7.62 × 39 <AK-47> and British .303 <Enfield>) are imported.[1, 2, 3]

## GENERAL INFORMATION ON BULLETS

On the order of 85–118 million pounds of lead is used each year in the production of bullets[4] in the United States.[5, 6] The exact number of each caliber and type of bullet (such as jacketed or hollow point) is not known, but some estimates of production volumes have been provided by the Sporting Arms and Ammunition Manufacturers' Institute[7] and are shown in Table 4.1. It is generally acknowledged that .22 caliber bullets are the dominant type sold. Table 4.2 provides some examples of typical bullet masses for various calibers. Using 70 grains (0.16 oz, 4.54 g) as an arbitrarily assumed average bullet mass allows the estimation that the 85–118 million pounds of bullet lead produces about 8.5–11.8 billion bullets per year in the United States.

## OVERVIEW OF BULLET PRODUCTION

Figure 4.1 is a simplified flow chart for bullet production and approximate mass of material involved in each of the processed materials. Table 4.3 has been prepared from the general information given in Figure 4.1 to illustrate the approximate number of bullets associated with each of the manufacturing steps or

[1]Greenberg, R. R. March 3, 2003. Verbal communication to committee after visiting the SHOT Show February 13–16, 2003.

[2]*Shotgun News* Special Interest Publications, Peoria, IL May 20, 2003. A collection of firearms related advertisements for retailers and wholesalers.

[3]CABL also has value for the matching of foreign-produced bullet lead; this value varies according to the lead's nation of origin and that nation's lead recycling and manufacturing processes. The analysis of foreign-produced bullets is not discussed in detail in this report.

[4]The committee assumes these numbers include lead for shot as well as bullets.

[5]Biviano, M. B.; Sullivan, D. E.; Wagner, L. A. *Total Materials Consumption: An Estimation Methodology and Example Using Lead—A Materials Flow Analysis.* USGS Circular: 1183. April, 1999. <http://pubs.usgs.gov/circ/1999/c1183>.

[6]Smith, G. R. USGS Minerals Yearbook 2001: Lead. Reston, VA 2001. <http://minerals.er.usgs.gov/minerals/pubs/commodity/lead/leadmyb01.pdf>.

[7]Green, K. D. *Introduction to the Bullet Manufacturing Process: Committee on Scientific Assessment of Bullet Lead Elemental Composition Comparison,* Washington, DC February 3, 2003.

TABLE 4.1 Annual Production of Ammunitions Produced in the United States

| Ammunition Type | No. Rounds Produced per Year, billions | No. Boxes Produced per Year, millions | No. Units per Box |
|---|---|---|---|
| Shotgun shells (all gauges) | 1.1 | 44 | 25 |
| Rifle, center fire | 0.25 | 12.5 | 20 |
| Pistol and revolver, center fire | 0.55 | 11 | 50 |
| Rifle and pistol, rimfire | 2 | 40 | 50 |

Source: See Footnote 7.

TABLE 4.2 Examples of Various Caliber and Style of Bullets and Estimated Bullet Mass

| | | Total Mass of Projectile (Mass of Pb if Jacketed) | | |
|---|---|---|---|---|
| Caliber | Style | Grains | Ounces | Grams |
| .22 Long rifle | Round nose/ Hollow point | 40 | 0.0914 | 2.59 |
| 9 × 19 mm | Lead round nose | 124 | 0.283 | 8.04 |
| 9 × 19 mm | Full metal jacket | 124 (103.0) | 0.283 (0.237) | 8.04 (6.71) |
| .38 special | Lead round nose | 150 | 0.343 | 9.72 |
| 44 Remington magnum | Lead truncated cone | 240 | 0.549 | 15.6 |
| 5.56 × 45 mm | Full metal jacket | 62 (31.6) | 0.142 (0.0722) | 4.02 (2.05) |
| 5.56 × 45 mm | Full metal jacket | 55 (46.1) | 0.126 (0.105) | 3.56 (2.99) |
| 7.62 × 51 mm | Full metal jacket | 145 (93.1) | 0.331 (0.213) | 9.40 (6.03) |

products. Calculations assumed a mass of 40 grains (0.0914 oz, 2.59 g) for a .22 rimfire projectile. The number of projectiles is based on 100 percent yield. Since some material is not converted directly to the final bullets (for example, initial piece of extruded wire, weep from bullet presses), the actual number of projectiles produced will be lower.

In the United States, secondary smelters melt recycled lead (primarily from recycled lead-acid storage batteries) for bullet lead processing in large pots.[8] The designation of *primary smelter* is reserved for manufacturing facilities that produce lead from ores. Such facilities are rarely associated directly with bullet production in the United States, but this is not the case in some foreign countries. Secondary smelting is reported to account for half the lead produced in the

[8]Smith, G. R. *Lead Recycling in the United States in 1998.* USGS Circular: 1196-F. 2002. <http://pubs.usgs.gov/circ/c1196f/>.

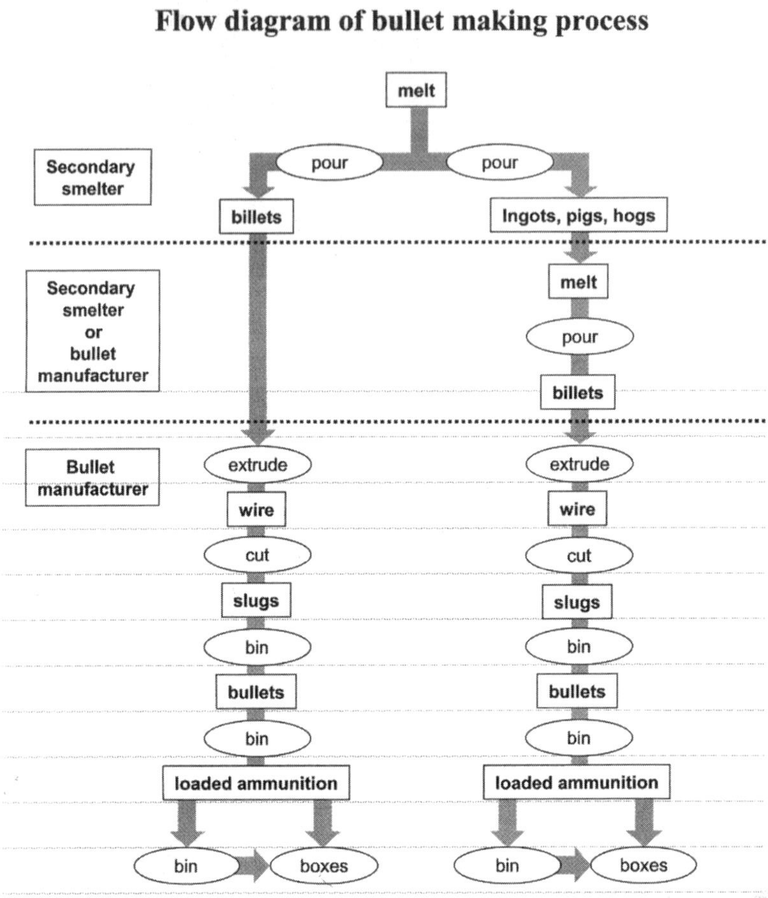


FIGURE 4.1 Flow diagram of bullet materials, a general description of the many steps involved in bullet production.

TABLE 4.3 Approximate Masses and Numbers of Bullets Produced from "Single Unit" of Various Stages in Manufacturing Process[a]

| Source of Material | Weight of Material (lbs) | Mass of Material (kg) | Yield (of .22 Caliber Bullets) |
|---|---|---|---|
| Melt pot | 200,000 | 90,719 | 35,000,000 |
| Melt pot | 100,000 | 45,360 | 17,500,000 |
| Sow | 2,000 | 907 | 350,000 |
| Billet | 70–350 | 32–159 | 12,250–61,250 |
| Pig/Ingot | 60–125 | 27–57 | 10,500–21,875 |

[a]Green, K. D. *Introduction to the Bullet Manufacturing Process: Committee on Scientific Assessment of Bullet Lead Elemental Composition Comparison*, Washington, DC February 3, 2003.

United States. There are 50 plants, with capacities ranging from 1,000 to 120,000 tons/year.[9]

Refining of the melt to remove various elements present either as impurities or as previously added alloy elements can occur at the secondary smelter.[10, 11] After refinement, Sb, less frequently Sn, and sometimes both elements may be added to harden the bullet. Finally, the melt is poured into various smaller products, including billets, which are sent to the bullet manufacturer.

The bullet manufacturer may use the purchased billets directly for production, but it is not uncommon for bullet manufacturers to remelt the purchased lead and cast their own billets for production.[12] The bullet manufacturer extrudes bullet wire from a solid billet; this results in one or more wires per billet, depending on whether the extruder die has one or more extrusion ports. Generally, a continuous wire is not produced from multiple billets due to the likelihood of discontinuity and the production of a flawed slug at the junction due to lead lamination. The size of the extruded wire is dictated by the caliber (diameter) of the bullet to be produced from that wire.

The bullet wire is then fed into a machine that cuts it to predetermined lengths to generate slugs of the approximate weight and dimensions of the final bullet. The slugs are collected in bins, whose size varies from plant to plant. In larger manufacturing facilities, several extruders may be operated in parallel in the production of slugs of a given caliber, and the slugs from the various extruders may be collected in the same bin. A given wire is converted to slugs of a given length and diameter.

The slugs are then pressed into the final shape of the bullet, a jacket is applied (if appropriate), and the bullets are again collected in bins.[13] The bullets are seated into appropriately prepared cartridge cases (loaded with primer and powder) to form the loaded ammunition, which is either collected in bins or sent directly to machinery for packing in boxes. The boxes generally contain 20–50 rounds each, depending on the caliber and the products being offered by the company. A more specific example of the wire-to-ammunition production steps for .22 caliber rimfire bullet production is as follows:[14]

---

[9]U.S. Environmental Protection Agency. Compilation of Air Pollutant Emission Factors, AP-42, Fifth Edition, Volume I: *Stationary Point and Area Scources,* Secondary Lead Chapter 12 section 11. Research Triangle Park, NC, January 1995.

[10]Randich, E.; Duerfeldt, W.; McLendon, W.; and Tobin, W. *Foren. Sci. Int.* **2002**, *127*, 174–191.

[11]Frost, G. E. *Ammunition Making*, Chapter 3. National Rifle Association of America, Washington, DC 1990, 25–43.

[12]Green, K. D. *Introduction to the Bullet Manufacturing Process: Committee on Scientific Assessment of Bullet Lead Elemental Composition Comparison*, Washington, DC February 3, 2003.

[13]Bullet cores are extruded as wires of a slightly smaller diameter than for unjacketed bullets of the same caliber, are cut into slugs, and are swaged into thimble-like jackets. The production of bullet cores is otherwise identical to the production of bullets.

[14]Green, K. D. *Introduction to the Bullet Manufacturing Process: Committee on Scientific Assessment of Bullet Lead Elemental Composition Comparison*, Washington, DC February 3, 2003.

- Pinch cut to partially perforate the wire.
- Tumble the partially perforated wire to break it into slugs.
- Swage press to final shape (three steps are needed).
- Wash and rinse.
- Flash plate with copper alloy (if high-velocity product is being made).
- Lubricate.
- Assemble into loaded ammunition.
- Pack ammunition in boxes.

The boxes are then generally bundled into appropriately sized shipping quantities—such as cartons, crates, or pallets—and sent to jobbers, distributors, wholesalers, or large retailers. They then go to the retailer's shelf for purchase by the consumer.

Reloaders, both commercial and private, are another source of loaded ammunition and are less directly connected to large-volume manufacturers.[15] Using refurbished cases for reloading, reloaders make less-expensive ammunition. In some instances, reloaded bullets are made from melted scrap lead, such as discarded wheel-balancing weights that are remelted and poured into bullet molds.

## DETAILS OF BULLET PRODUCTION

This section details the various stages leading to the production and distribution of boxes of loaded ammunition. Comments on the variations that are known to exist at various stages are given here, but their implications for the homogeneity of melts, billets, wires, and so on, are discussed in the section titled "Compositional Information."

### Sources and Use of Lead

With over 3.5 billion pounds of lead smelted each year in the Unites States, the 85–118 million pounds used in bullet manufacturing comprises about 2.5–3 percent of total lead use; lead-acid storage batteries probably represent the largest product.[16, 17] Secondary smelters that produce bullet lead are also gen-

[15]Commercial reloaders are often known as remanufacturers. The concentrations of elements in component bullets used by reloaders are similar to the concentrations in bullet lead used by major manufacturers. Component bullet unit sales are a small fraction (5–10 percent) of loaded ammunition sales, but can follow wider distribution channels because there are fewer shipping restrictions. Reloaded ammunition is not expected to comprise a large percentage of the ammunition involved in casework.

[16]Biviano, M. B.; Sullivan, D. E.; Wagner, L. A. *Total Materials Consumption: An Estimation Methodology and Example Using Lead—A Materials Flow Analysis.* USGS Circular: 1183. April, 1999. <http://pubs.usgs.gov/circ/1999/c1183>.

[17]Smith, G. R. *USGS Minerals Yearbook 2001: Lead.* Reston, VA 2001. <http://minerals.er.usgs.gov/minerals/pubs/commodity/lead/leadmyb01.pdf>.

erally involved in the production of "battery lead." Chemical compositional requirements for bullet lead are much less stringent (that is, they have less-restrictive tolerances) than are needed for battery lead. However, a hardened lead is generally needed for bullets.[18, 19] Hardening is typically accomplished by the addition of Sb to the melt. Sn can also be used, but it is more expensive. Other components of bullet lead are generally carried over from the lead source, and maximal tolerances in their concentrations are normally specified by the bullet manufacturer.

Bullets are reportedly produced mainly from recycled lead in the United States. Therefore, it is impossible to trace bullet lead back to the original source of the ore,[20] and no detailed discussion will be presented here on the primary smelters and ore processing except to note that the ores are sulfides and contain small amounts of Cu, Fe, Zn, precious metals, and other trace and minor elements, such as As, Sb, and Bi. The primary smelting process involves removal of those elements by reduction and refining.

## Secondary Lead Smelters

As noted previously, the dominant source of bullet lead is the electrode materials from recycled batteries. The melting process takes place in pots that may contain, for example, 50–350 tons of melt. The descriptions given below are typical; they might not be applicable to all smelters.

The first step in secondary lead refining is treatment of scrap to remove metallic and nonmetallic contaminants. That is done by mechanical breaking and crushing to separate extraneous contaminants and then "sweating" the separated lead scrap in a reverberatory furnace to isolate the lead from metals that have higher melting points. The next step is smelting in a blast furnace to make "hard" (high-Sb) lead or in a reverberatory furnace to make "semisoft" (3–4 percent Sb) lead. Refining is normally done in a batch process that takes a few hours to a few days in kettle-type furnaces that have production capacities of 25–150 tons/day.[21] In the refining process, Cu, Sb, As, and Ni are the main elements removed. It is generally assumed that Sb is the element whose content is most critical because it determines the bullet hardness.[22, 23]

---

[18]Randich, E.; Duerfeldt, W.; McLendon, W.; and Tobin, W. *Foren. Sci. Int.* **2002,** *127,* 174–191.

[19]Peters, C.; Havekost, D. G.; Koons, R. D. *Crime Lab. Digest* **1988**, *15(2)*, 33–38.

[20]Smith, G. R. *USGS Minerals Yearbook 2001: Lead.* Reston, VA 2001. <http://minerals.er.usgs.gov/minerals/pubs/commodity/lead/leadmyb01.pdf>.

[21]U.S. Environmental Protection Agency. Compilation of Air Pollutant Emission Factors, AP-42, Fifth Edition, Volume I: *Stationary Point and Area Sc=ources*, Secondary Lead Chapter 12 section 11. Research Triangle Park, NC, January 1995.

[22]Randich, E.; Duerfeldt, W.; McLendon, W.; and Tobin, W. *Foren. Sci. Int.* **2002**, *127*, 174–191.

[23]Peters, C.; Havekost, D. G.; and Koons, R. D. *Crime Lab. Digest* **1988**, *15(2)*, 33–38.

TABLE 4.4 Example of Manufacturer's Compositional Requirements for Pb to Be Used in .22 Long Rifle Projectiles[a]

| Preferred Analysis | Weight Percent |
|---|---|
| Sb | 0.85 ± 0.15 % |
| **Maximal Impurities** | **Weight Percent** |
| Al | 0.001% |
| As | 0.05–0.10% |
| Bi | 0.05% |
| Cd | 0.001% |
| Cu | 0.03% |
| Ca | 0.001% |
| Fe | 0.001% |
| Ni | 0.001% |
| Se | 0.002% |
| Ag | 0.01% |
| S | 0.001% |
| Te | 0.01% |
| Sn | 0.15–0.2% |
| Zn | 0.001% |
| **Sow Size** | **Weight in Pounds** |
| Maximum | 2,200 lb |
| Minimum | 1,500 lb |

[a]Prengaman, R. D. *Lead and Lead Refining: Committee on Scientific Assessment of Bullet Lead Elemental Composition Comparison*, Washington, DC March 3, 2003.

In the production of bullet lead, the manufacturer generally has requirements for the concentrations of the final lead alloy.[24] The elemental compositional requirements can vary with the bullet manufacturer. Depending on the element, either maximal allowable or ranges of concentrations may be specified. Table 4.4 shows an example of one manufacturer's compositional requirements for lead to be used in .22 long rifle bullets. Some bullet producers use as-received billets from secondary smelters, and others conduct tertiary melting to make additional adjustments to the lead composition or to recycle scraps of lead produced during bullet production.

A secondary smelter may produce solid lead of various shapes, including ingots, pigs, and billets. An analysis certificate accompanies the product shipped to the bullet manufacturer; it uses a smelter-dependent format that contains various degrees of analytical detail. Spark-emission optical spectroscopy is the technique generally used for analysis of the alloy at the smelters.[25] The technique

[24]Prengaman, R. D. *Lead and Lead Refining: Committee on Scientific Assessment of Bullet Lead Elemental Composition Comparison*, Washington, DC March 3, 2003.

[25]Prengaman, R. D. *Lead and Lead Refining: Committee on Scientific Assessment of Bullet Lead Elemental Composition Comparison*, Washington, DC March 3, 2003.

generally produces precision on the order of ±10–20%; however, when the most stringent standardization procedures are implemented, precision may approach ±5 percent.[26]

There is no requirement by the bullet manufacturers that all lead ingots received from a smelter come from a single pour or melt. It is generally assumed that the composition of a given melt is constant and homogeneous from the beginning to the end of the pour if nothing is added to the pot during the pour.[27] The assumption of homogeneity is based on the convective mixing in the vat and the relatively short pouring time. It should be noted that during a pour material may be added to the original melt, thus producing time-varying compositional changes. The additions may include bulk material (ingots, pigs, and so on), manufacturing scrap (pieces of bullet wire, scrap from bullet-forming operations, and the like), or molten lead introduced from a secondary vat. Examples of the time-dependent variation in composition can be seen in some of the data of Koons and Grant.[28] In the case of at least one manufacturer, billets are not poured from a vat that has a constant composition; instead, while the vat is being poured, molten lead from another pot is continuously added to maintain the level of molten lead in the vat being poured. Thus, compositional changes can occur during casting. The data of Koons and Grant[29] indicate that compositional change occurs over several 60 lb ingots that were being poured. For example, the concentration of Sn decreased by 60 percent (from 0.030 to 0.012 percent Sn) over a 30 minute period, the largest change of the data presented. Combining this information with the standard deviations for the analytical measurement (that is, < 0.001 percent Sn) it can be estimated that approximately 15 ingots (approximately 850 lbs of Pb) were poured before the average concentrations changed by one standard deviation. Thus, it can be reasonably assumed that the rate of compositional change—even when molten lead batches are mixed during a pour—from one poured ingot to the next poured ingot is much smaller than the measurement precision available. It also follows that any compositional change in the lead initially poured into an ingot (or billet) would be indistinguishable from the molten lead added to the mold to complete the pour of that ingot, as long as the casting of the ingot was completed in a single pour.

Randich et al.[30] also showed occasional distinct concentration changes in some elements as samples were extracted from the beginning, middle, and end of the pour. Statistical analysis of the changes showed that there was no distinct time-dependent one-directional change (that is, always increasing or decreasing

---

[26]Mitteldorf, A. J. In *Trace Analysis*; Morrison, G. H., Ed.; John Wiley and Sons: New York, 1965, pp 193–243.

[27]Prengaman, R. D. *Lead and Lead Refining: Committee on Scientific Assessment of Bullet Lead Elemental Composition Comparison*, Washington, DC March 3, 2003.

[28]Koons, R. D. and Grant, D. M. *J. Foren. Sci.* **2002**, *47*, 950–958.

[29]Koons, R. D. and Grant, D. M. *J. Foren. Sci.* **2002**, *47*, 950–958.

[30]Randich, E.; Duerfeldt, W.; McLendon, W.; and Tobin, W. *Foren. Sci. Int.* **2002**, *127*, 174–191.

as the pour proceeded), which would suggest for these data that lead of a different composition was being added during the pour, rather than that some chemical process occurred that depleted or enriched a given element as a function of time. The former possibility (the addition of lead during the pour) is supported by the data of Koons and Grant,[31] who presented a more detailed analysis of billets resulting from pours. Koons and Grant used several of the same data sets as Randich et al.[32]

## Billet Production

Billets weigh 70–350 lbs (32–159 kg), depending on the manufacturer and the size and type of extruder that is used in the production of bullet wire.[33, 34] In some instances, the secondary smelter is also a bullet manufacturer, and the billets produced are used on site in the production of wire, slugs, and so forth. In other instances, the lead ingots, pigs, or billets are shipped to bullet manufacturers, and the bullet manufacturers may use the billets directly in their extruders to produce wire. There are also instances in which the ingots or pigs obtained from the secondary smelters are remelted to pour new billets at the bullet manufacturing plant.

Various activities can occur during this tertiary melting that affect the final billet composition. For example, melted lead prior to casting in billets is typically "fluxed" to remove oxidized lead metal elements and other impurities. The fluxing agent can contain a number of different materials, and is often borate-based in commercial bullet manufacturing operations. Nitrogen gas is also a common fluxing agent. The flux entrains the impurities and floats them to the surface of the lead melt for removal.

## Bullet Production

Billets are used without alteration (in their original, solid state) in the extruders to produce bullet wire. The mass of the wire is somewhat less than the mass of the billet, because the tail end of the billet cannot be forced through the extrusion die by the ram.[35, 36] The length of the wire is governed by the billet

---

[31]Koons, R. D. and Grant, D. M. *J. Foren. Sci.* **2002**, *47*, 950–958.

[32]Randich, E.; Duerfeldt, W.; McLendon, W.; and Tobin, W. *Foren. Sci. Int.* **2002**, *127*, 174–191.

[33]Green, K. D. *Introduction to the Bullet Manufacturing Process: Committee on Scientific Assessment of Bullet Lead Elemental Composition Comparison*, Washington, DC February 3, 2003.

[34]Prengaman, R. D. *Lead and Lead Refining: Committee on Scientific Assessment of Bullet Lead Elemental Composition Comparison*, Washington, DC March 3, 2003.

[35]Green, K. D. *Introduction to the Bullet Manufacturing Process: Committee on Scientific Assessment of Bullet Lead Elemental Composition Comparison*, Washington, DC February 3, 2003.

[36]Prengaman, R. D. *Lead and Lead Refining: Committee on Scientific Assessment of Bullet Lead Elemental Composition Comparison*, Washington, DC March 3, 2003.

size and the wire diameter (bullet caliber). For example, a 70-lb billet should produce about 114 ft of wire intended for .22 caliber ammunition, but the same billet should produce about 27 ft of wire if .45 caliber bullets are the intended product. The extruder die may have a single exit port that produces a single wire strand from the billet, or it may have multiple extrusion ports that produce several wires from a single billet. Several feet of the wire formed at the beginning of the extrusion process may be discarded and recycled into a future billet.[37]

In brief, the wire is used as feed for a cutter, which consists of a machine that automatically introduces the wire into a cutting device to produce slugs, small cylinders of lead whose length and mass are close to those of the final bullet. The slugs are stored in large bins that may hold substantial quantities of slugs from different wires.

The binned slugs are fed into hoppers that feed the presses that form the bullets. Although it is not a true swaging process, this term is commonly encountered in the literature describing the process. Thus formed, the bullets are then tumbled, sometimes lubricated, and stored in bins.[38, 39] For some bullet types, a metal jacket is added.

## Production of Loaded Ammunition

The loaded ammunition, which is sometimes referred to as rounds or cartridges, consists of a brass case that is charged with primer and powder and into which the bullet is pressed. Bullets and cases from bins are fed into hoppers, and the process of ammunition production proceeds in an automated fabrication machine. The product is sent directly to the packaging operation or is placed in large bins for later packaging.[40, 41]

## Packaging and Distribution

The bullet manufacturer packages the ammunition in boxes for shipment. The box typically is labeled with a stamp that refers to the "boxing lot," which may be recorded as a date or simply a number. In some manufacturing plants,

[37]Frost, G. E. *Ammunition Making*, Chapter 3. National Rifle Association of America, Washington, DC 1990, 25–43.

[38]Frost, G. E. *Ammunition Making*, Chapter 3. National Rifle Association of America, Washington, DC 1990, 25–43.

[39]In some cases, bullets may be washed, rinsed, and plated in addition to being tumbled and lubricated. Each step can introduce further mixing of bullets from different lead wires and discrete sections of lead wire.

[40]Green, K. D. *Introduction to the Bullet Manufacturing Process: Committee on Scientific Assessment of Bullet Lead Elemental Composition Comparison*, Washington, DC February 3, 2003.

[41]Frost, G. E. *Ammunition Making*, Chapter 3. National Rifle Association of America, Washington, DC 1990, 25–43.

the boxing lot number refers to the date the ammunition was loaded; in others, the date or number is not necessarily related to a particular stage in the production process. A typical box contains 20–50 cartridges, but some units or boxes are larger, depending on product line and caliber. For example, .22 long rifle "value packs" are commonly sold in 550-round boxes, and 100-round boxes of 9 × 19 mm ammunition have recently become common at larger retailers.[42] The boxes are arranged in larger shipping units (such as cartons, crates, and pallets) and shipped to jobbers, distributors, wholesalers, or large retailers.

Attempts to obtain details on the shipping and distribution processes for loaded ammunition were unsuccessful and therefore are not clearly understood by the committee. For example, the committee has no evidence that distribution from a given manufacturer is regional as has been suggested in one report.[43] Similarly, the frequency and size of shipments are unknown, but they are expected to vary widely, depending on the customer and the type of ammunition. However, it is reasonable to assume that high-turnover ammunition (for example, .22 caliber) is shipped more frequently than others and in larger quantities.

The committee has a similar lack of knowledge about retail dispersion of boxes. For example, it is not known whether first-in-first-out sales occur—that is, whether older shipments are arranged on shelves to be sold first.

## COMPOSITIONAL INFORMATION

Multiple steps are required to move from bullet production to boxes of ammunition, and manufacturers vary in their processing of materials leading to bullet formation. In addition, storage times before actual packaging and shipping depend heavily on caliber; for example, high-production munitions, such as .22 caliber, probably move more rapidly from slug production to shipping than less-common munitions.

### Homogeneity

There is much debate of the homogeneity of the lead "source." It is unclear whether macro- and microscale inhomogeneities are present at some or all of the stages of lead and bullet production and if such inhomogeneities would affect CABL. The poor definition and understanding of the term "source" causes additional confusion. These topics are clarified below.

- *Melt.* It is reasonable to assume that a given batch of molten lead exhibits sufficient mixing (such as convective stirring because of the heating process)

[42]*Shotgun News* Special Interest Publications, Peoria, IL May 20, 2003. A collection of firearms related advertisements for retailers and wholesalers.

[43]Randich, E.; Duerfeldt, W.; McLendon, W.; and Tobin, W. *Foren. Sci. Int.* **2002**, *127*, 174–191.

for compositional homogeneity to develop quickly in the melt, assuming that there are no additions to the molten vat during pouring. Some constituents—such as Sb, As, and Sn—oxidize in air, and their loss or flotation to the surface is expected to take place slowly. However, the rate of compositional change is unlikely to be significant relative either to the rate of casting of billets or to the uncertainty of the concentrations of these materials. The assumption that the rate of compositional change is insignificant is supported by the small surface area exposed to air relative to the total mass of the melt.

- *Pigs, Ingots, and Billets.* The homogeneity of ingots, pigs, and other large blocks of smelted lead is not an issue, because they are always remelted before billets are cast. Inhomogeneity of billets can arise from two factors. First, a billet may be cast in two stages, with the second stage long enough after the first for a measurable compositional difference to exist, depending on the constancy of the melt between the two pours that finalize billet production. Second, solutes inevitably segregate to the center of the billet during solidification.
- *Wires, Slugs, and Bullets.* The extrusion process used to produce the wire from a billet is thought to negate the inhomogeneity due to segregation during solidification because the flow of the solid is turbulent as the billet enters the mouth of the die. Uniformity along the length of wire has not been substantiated. However, Koons and Grant have sampled wires produced from billets from a pour and found that concentrations remained constant (that is, within analytical precision) over several billets.[44] Small compositional differences *may* exist along the length of the wire as a result of several factors. Segregation of material at the end of the billet mold may enrich the less refractory constituents in the lead, and detectable segregation will diminish as the impurity level decreases. If this segregation occurs, it still might not contribute to compositional differences along the length, because several feet of the first length of wire extruded are discarded and returned to a scrap bin. If multiple billets are loaded into an extruder, a continuous, *single wire* is extruded, but is cut into separate wires where the change of billets takes place.[45] It is not clear from the data available whether the concentration of Sb is segregated in the billet or wire. While a paucity of data also exists for the spatial dependence of concentration of the other impurities along the length of wire (or in the billet), their significantly lower concentration should make spatial inhomogeneities less likely. It is reasonable to assume that cutting the wire to produce the slugs and pressing the slugs to form the final bullets produce no substantial segregation of elements in the lead.
- *Mixing of Slugs, Bullets, and Loaded Ammunition.* Some manufacturers

[44]Koons, R. D. and Grant, D. M. *J. Foren. Sci.* **2002**, *47*, 950–958.

[45]Frost, G. E. *Ammunition Making*, Chapter 3. National Rifle Association of America, Washington, DC 1990, 25–43.

use multiple cutting machines with distinct wire feeds to simultaneously produce slugs that are collected in a common slug bin. Similarly, a given production run may require sequential cutting of several wires and collection in a common bin. Thus, if wires are not of the same composition, a bin can contain slugs with a finite number of distinct compositions and if slugs from previous runs went unused at the start of the cutting of new wires, they contribute to the mixing of slugs of different compositions in a bin.

The slug bins are emptied into hoppers that feed the bullet-shaping presses, and the bullets formed may be collected in bullet bins before they are fitted into cases to form loaded ammunition. "Tail-in-tail-out" mixing can occur in the bins if their full contents are not used in a single production run of ammunition. The mixing with previously formed bullets will not occur if the pressed bullets are used immediately (without storage in bullet bins) in ammunition production.

The loaded ammunition can be routed directly to a packaging area, in which case no additional mixing occurs. However, loaded ammunition is sometimes stored temporarily in ammunition bins, where batch mixing and tail-in-tail-out procedures that contribute to mixing can occur.

The likelihood of mixing in the various bins described above is supported by the compositional analyses conducted on the bullets in a given box of ammunition.[46] It is routinely found that a single box contains multiple distinct compositional groupings—as many as 14.[47]

- *Boxes, Crates, and Distribution.* The boxes of ammunition are generally stamped with a box lot number. Depending on the manufacturer, this lot number may only reflect the packaging date, may be a direct indication of the date and shift during which the ammunition was loaded, or may be a code indicating packing date and shift, which can be traced through the manufacturer's internal records to one or more shifts of loading operations. A stamped date does not reflect the date of pouring of billets, extrusion of wire, or formation of bullets. If filled boxes are stored on shelves because of overruns, boxes of different runs (with different dates) may be mixed in larger shipping units. Thus, a large-volume shipping unit for more commonly used ammunition might or might not contain only boxes with the same lot number and date.

As noted previously, distribution of boxes, crates, pallets, and other quantities of ammunition is poorly understood; there is minimal documentation to assist in establishing general trends. It is clear that distribution can lead to varied scenarios regarding retail dispersion of bullets from a distinct compositional group.

---

[46]Peters, C.; Havekost, D. G.; Koons, R. D. *Crime Lab. Digest* **1988**, *15(2)*, 33–38.

[47]Peele, E. R.; Havekost, D. G.; Peters, C. A.; Riley, J. P.; Halberstam, R. C.; and Koons, R. D. In *Proceedings of the International Symposium on the Forensic Aspects of Trace Evidence* June 24–28, 1991, pp 57–68.

### THE "SOURCE"

When the metal compositions of two bullets are analytically indistinguishable, it is commonly suggested that they may have originated in the same "source." It might be good to replace that vague term with "compositionally indistinguishable volume of lead" (CIVL). The CIVL, *produced during one production run at one point in time,* is at least as large as the sample taken for analysis. From the current understanding of the bullet production process, CIVL can refer to different tangible products associated with the manufacturing cycle. At its largest, the CIVL may be a vat of molten lead whose composition is not altered during the pouring of billets. Similarly, the CIVL may consist of a series of billets that were poured before the vat composition was altered by, for example, the addition of more molten lead to replenish the vat. At the very least, a CIVL may consist of several wires. The ramifications of identifying bullets whose compositions are analytically indistinguishable and their possible association with a single CIVL are discussed later in this chapter.

## COMPOSITIONAL ANALYSIS OF BULLET LEAD AS EVIDENCE IN THE LEGAL SYSTEM

This section discusses the legal aspects of CABL evidence. Knowledge of the lead and bullet manufacturing processes underlies the proper interpretation of CABL evidence. The topics covered here include admissibility standards (including evaluation of match data) and pretrial discovery.

### ADMISSIBILITY STANDARDS

The admissibility of CABL raises issues concerning expert testimony and relevance.

#### Expert Testimony

Experts are called by the prosecution to testify to the fact of matching and, in most cases, the evidentiary implication of a match. Federal Rule of Evidence 702 governs the admissibility of expert testimony in federal trials:

> If scientific, technical, or other specialized knowledge will assist the trier of fact to understand the evidence or to determine a fact in issue, a witness qualified as an expert by knowledge, skill, experience, training, or education, may testify thereto in the form of an opinion or otherwise, if (1) the testimony is based upon sufficient facts or data, (2) the testimony is the product of reliable principles and methods, and (3) the witness has applied the principles and methods reliably to the facts of the case.

In *Daubert v. Merrell Dow Pharmaceuticals, Inc.*,[48] the Supreme Court interpreted an earlier version of Rule 702 to require that scientific evidence meet a *reliability* test. The Court wrote that "in order to qualify as 'scientific knowledge,' an inference or assertion must be derived by the scientific method. Proposed testimony must be supported by appropriate validation—i.e., 'good grounds,' based on what is known. In short, the requirement that an expert's testimony pertain to 'scientific knowledge' establishes a standard of evidentiary reliability."[49] The Court held that the *Frye* test,[50] which required that a novel scientific technique be generally accepted in the relevant scientific community as the sole condition for admissibility,[51] had been superseded by Rule 702 of the Federal Rules of Evidence.

Under the *Daubert* analysis, the trial court must make "a preliminary assessment of whether the reasoning or methodology underlying the testimony is scientifically valid and of whether that reasoning or methodology properly can be applied to the facts in issue."[52] In performing this "gatekeeping function," the trial court may consider a number of factors: whether the theory or technique can be and has been tested,[53] whether it has been subjected to peer review and

---

[48]509 U.S. 579 (1993). *See* Margaret A. Berger, *The Supreme Court's Trilogy on the Admissibility of Expert Testimony*, in Federal Judicial Center, Reference Manual on Scientific Evidence 9 (2d ed. 2000); David L. Faigman et al., Modern Scientific Evidence ch. 1 (2d ed. 2002); 1 Paul C. Giannelli & Edward J. Imwinkelried, Scientific Evidence ch. 1 (3d ed. 1999).

[49]509 U.S. at 590. The Court also commented that "under the Rules the trial judge must ensure that any and all scientific testimony or evidence admitted is not only relevant, but reliable." *Id.* at 589. "In short, the requirement that an expert's testimony pertain to 'scientific knowledge' establishes a standard of evidentiary reliability." *Id.* at 590. In footnote 9, the Court elaborated: "We note that scientists typically distinguish between 'validity' (does the principle support what it purports to show?) and 'reliability' (does application of the principle produce consistent results?). . . . [O]ur reference here is to *evidentiary* reliability—that is, trustworthiness. . . . In a case involving scientific evidence, *evidentiary reliability* will be based upon scientific validity."

[50]Frye v. United States, 293 F. 1013, 1014 (D.C. Cir. 1923). *See* Paul C. Giannelli, *The Admissibility of Novel Scientific Evidence: Frye v. United States, a Half-Century Later,* 80 Columbia L. Rev. 1197 (1980).

[51]As noted below, "general acceptance" continues as a factor under *Daubert* but not the sole criterion for admissibility as under *Frye*.

[52]509 U.S. at 592-93. In a later passage, the Court wrote that "the Rules of Evidence—especially Rule 702—do assign to the trial judge the task of ensuring that an expert's testimony both rests on a reliable foundation and is relevant to the task at hand. Pertinent evidence based on scientifically valid principles will satisfy those demands." *Id.* at 597. *See also* Fed. R. Evid. 104(a) ("Preliminary questions concerning . . . the admissibility of evidence shall be determined by the court . . . .").

[53]*Id.* at 593 ("Ordinarily, a key question to be answered in determining whether a theory or technique is scientific knowledge that will assist the trier of fact will be whether it can be (and has been) tested. 'Scientific methodology today is based on generating hypotheses and testing them to see if they can be falsified; indeed, this methodology is what distinguishes science from other fields of human inquiry.' Green 645. See also C. Hempel, Philosophy of Natural Science 49 (1966) ('[T]he statements constituting a scientific explanation must be capable of empirical test'); K. Popper, Conjectures and Refutations: The Growth of Scientific Knowledge 37 (5th ed. 1989) ('[T]he criterion of the scientific status of a theory is its falsifiability, or refutability, or testability') (emphasis deleted).").

publication,[54] a technique's known or potential error rate, the existence and maintenance of standards controlling the technique's operation,[55] and a technique's general acceptance in the relevant scientific community.[56] Those factors, however, are neither dispositive nor exhaustive. The Court emphasized that the Rule 702 standard is "a flexible one."

The Court followed with *General Electric Co. v. Joiner*[57] and *Kumho Tire Co. v. Carmichael*[58] to make up what is now known as the *Daubert* trilogy. *Daubert* and its progeny have come to be viewed as establishing a stringent standard of admissibility.[59] In *Weisgram v. Marley Co.*,[60] the Supreme Court remarked: "Since *Daubert*, . . . parties relying on expert evidence have had notice of the exacting standards of reliability such evidence must meet."[61] More-

[54]*Id.* 593-94 ("Another pertinent consideration is whether the theory or technique has been subjected to peer review and publication. Publication (which is but one element of peer review) is not a *sine qua non* of admissibility; it does not necessarily correlate with reliability, and in some instances well-grounded but innovative theories will not have been published. Some propositions, moreover, are too particular, too new, or of too limited interest to be published. But submission to the scrutiny of the scientific community is a component of 'good science,' in part because it increases the likelihood that substantive flaws in methodology will be detected. The fact of publication (or lack thereof) in a peer reviewed journal thus will be a relevant, though not dispositive, consideration in assessing the scientific validity of a particular technique or methodology on which an opinion is premised.") (citations omitted).

[55]*Id.* at 594.

[56]*Id.* ("Widespread acceptance can be an important factor in ruling particular evidence admissible, and 'a known technique which has been able to attract only minimal support within the community,' . . . may properly be viewed with skepticism.").

[57]522 U.S. 136 (1997) (specifying that the admissibility decision is to be reviewed on appeal under an abuse-of-discretion standard).

[58]526 U.S. 137 (1999). In *Kumho*, the Court extended *Daubert's* reliability requirement to nonscientific expert testimony under Rule 702: "*Daubert's* general holding—setting forth the trial judge's general 'gatekeeping' obligation—applies not only to testimony based on 'scientific' knowledge, but also to testimony based on 'technical' and 'other specialized' knowledge." *Id.* at 141.

[59]*See* Rider v. Sandoz Pharm. Corp., 295 F.3d 1194, 1202 (11th Cir. 2002) ("The district court, after finding that the plaintiffs' evidence was unreliable, noted that certain types of other evidence may have been considered reliable, including peer-reviewed epidemiological literature, a predictable chemical mechanism, general acceptance in learned treatises, or a very large number of case reports."); Jerome P. Kassirer and Joe S. Cecil, *Inconsistency in Evidentiary Standards for Medical Testimony: Disorder in the Courts*, 288 J. Am. Med. Assn. 1382, 1382 (2002) ("In some instances, judges have excluded medical testimony on cause-and-effect relationships unless it is based on published, peer-reviewed, epidemiologically sound studies, even though practitioners rely on other evidence of causality in making clinical decisions, when such studies are not available.").

[60]528 U.S. 440 (2000) (reviewing a summary judgment in a wrongful death action against a manufacturer of an allegedly defective baseboard heater).

[61]*Id.* at 455. *See also* Brooke Group, Ltd. v. Brown & Williamson Tobacco Corp., 509 U.S. 209, 242 (1993) ("When an expert opinion is not supported by sufficient facts to validate it in the eyes of the law, or when indisputable record facts contradict or otherwise render the opinion unreasonable, it cannot support a jury's verdict.").

over, some federal courts have read the *Daubert* trilogy as inviting a "reexamination even of 'generally accepted' venerable, technical fields."[62]

In 2000, Rule 702 was amended[63] to codify *Daubert* and *Kumho*.[64] The Advisory (drafting) Committee's note to that rule supplements the *Daubert* factors with other considerations: whether the underlying research was conducted independently of litigation, whether the expert unjustifiably extrapolated from an accepted premise to an unfounded conclusion, whether the expert has adequately accounted for obvious alternative explanations, whether the expert was as careful as he or she would be in professional work outside of paid litigation, and whether the field of expertise claimed by the expert is known to reach reliable results.[65]

The *Daubert* decision is restricted to federal trials; it does not apply to other jurisdictions.[66] Thus, states are free to determine their own standards for admissibility of expert testimony, even in the 40 or so jurisdictions that have adopted evidence rules based on the Federal Rules of Evidence. Many jurisdictions have adopted the *Daubert* framework.[67] Moreover, other jurisdictions had rejected the *Frye* test before the *Daubert* decision,[68] and many of these now look to *Daubert* for guidance.[69]

---

[62]United States v. Hines, 55 F. Supp. 2d 62, 67 (D. Mass. 1999). *See also* United States v. Hidalgo, 229 F. Supp. 2d 961, 966 (D. Ariz. 2002) ("Courts are now confronting challenges to testimony . . . whose admissibility had long been settled."). Nevertheless, other courts seem to apply a less stringent approach to some long accepted forensic techniques. *See* United States v. Crisp, 324 F.3d 261 (4th Cir. 2003) (fingerprint and handwriting comparison; compare majority and dissenting opinions).

[63]The following clause was added to Rule 702: "if (1) the testimony is based upon sufficient facts or data, (2) the testimony is the product of reliable principles and methods, and (3) the witness has applied the principles and methods reliably to the facts of the case."

[64]Some courts believe the amendment went beyond *Daubert* and *Kumho*. *See* Rudd v. General Motors Corp., 127 F. Supp. 2d 1330, 1336-37 (M.D. Ala. 2001) ("[T]he new Rule 702 appears to require a trial judge to make an evaluation that delves more into the facts than was recommended in *Daubert*, including as the rule does an inquiry into the sufficiency of the testimony's basis ('the testimony is based upon sufficient facts or data') and an inquiry into the application of a methodology to the facts ('the witness has applied the principles and methods reliably to the facts of the case'). Neither of these two latter questions that are now *mandatory* under the new rule . . . were expressly part of the former admissibility analysis under *Daubert*.").

[65]Fed. R. Evid. 702 advisory committee's note (2000).

[66]*Daubert*, 509 U.S. at 587 ("We interpret the legislatively enacted Federal Rules of Evidence as we would any statute.").

[67]Alaska, Colorado, Connecticut, Idaho, Indiana, Kentucky, Massachusetts, Nebraska, New Hampshire, New Mexico, Oklahoma, South Dakota, Tennessee, and West Virginia. *See* 1 Paul C. Giannelli & Edward J. Imwinkelried, Scientific Evidence § 1-13 (3d ed. 1999).

[68]Arkansas, Delaware, Georgia, Iowa, Montana, North Carolina, Ohio, Oregon, Rhode Island, South Carolina, Texas, Utah, Vermont, and Wyoming. *See id.* at § 1-14.

[69]*E.g.*, Nelson v. State, 628 A.2d 69, 73 (Del. 1993) ("Our decisions [in prior cases] are consistent with the Supreme Court's decision in *Daubert*."); State v. Foret, 628 So. 2d 1116, 1123 (La. 1993) ("Past decisions of this court have espoused similar sentiments [as *Daubert*] . . .").

Nevertheless, some jurisdictions have retained the *Frye* rule.[70] Because Federal Bureau of Investigation (FBI) examiners testify in state trials, the *Frye* general-acceptance standard may apply to CABL in some cases.[71]

## Relevance and Its Counterweights

Relevance is the threshold issue for all evidence. Federal Rule 401 defines *relevant evidence* as "evidence having any tendency to make the existence of [a material or consequential fact] more probable or less probable than it would be without the evidence." Rule 401's standard does *not* require that the evidence make a consequential (material) fact "more probable than not" ("preponderance of evidence") but only that the material fact (for example, the identity of a perpetrator) be more probable or less probable *with the evidence than without the evidence.*[72]

Rule 402 makes relevant evidence admissible in the absence of a rule of exclusion, and Rule 403 specifies circumstances under which a trial court is permitted to exclude relevant evidence. Rule 403 reads: "Although relevant, evidence may be excluded if its probative value is substantially outweighed by the danger of unfair prejudice, confusion of the issues, or misleading the jury, or by considerations of undue delay, waste of time, or needless presentation of cumulative evidence." In *Daubert*, the Supreme Court noted that "expert evi-

---

[70]*E.g.*, People v. Leahy, 882 P.2d 321, 323 (Cal. 1994) (The "*Kelly* formulation [of *Frye* under the Cal. Evid. Code] survived *Daubert*. . . ."); People v. Miller, 670 N.E.2d 721, 731 (Ill. 1996) ("Illinois follows the *Frye* standard for the admission of novel scientific evidence."); Burral v. State, 724 A.2d 65, 80 (Md. 1999) (Despite *Daubert*, "we have not abandoned *Frye* or *Reed*."). Other *Frye* jurisdictions include Alabama, Arizona, Florida, Kansas, Michigan, Minnesota, Mississippi, Missouri, Nevada, New Jersey, New York, Pennsylvania, and Washington. *See* 1 Giannelli & Imwinkelried, Scientific Evidence § 1-15 (3d ed. 1999).

[71]Some jurisdictions adhere to a third approach, known as the relevance approach. *See* State v. Peters, 534 N.W.2d 867, 873 (Wis. Ct. App. 1995) ("Once the relevancy of the evidence is established and the witness is qualified as an expert, the reliability of the evidence is a weight and credibility issue for the fact finder and any reliability challenges must be made through cross-examination or by other means of impeachment."); State v. Donner, 531 N.W.2d 369, 374 (Wis. Ct. App. 1995) ("[B]efore *Daubert*, the *Frye* test was not the law in Wisconsin. To that extent, Wisconsin law and *Daubert* coincide. Beyond that, Wisconsin law holds that 'any relevant conclusions which are supported by a qualified witness should be received unless there are other reasons for exclusion.' Stated otherwise, expert testimony is admissible in Wisconsin if relevant and will be excluded only if the testimony is superfluous or a waste of time. . . . Assuming that *Daubert* in its application represents something beyond *Walstad*, we observe that we . . . are bound to follow our supreme court case law.") (citations omitted).

[72]In some situations, the relevance of evidence depends on science—or at least knowledge outside the common experience of laypersons. *See* Fed. R. Evid. 401 advisory committee's note (federal drafters noted that relevance decision are based on "experience or science, applied logically to the situation at hand").

dence can be both powerful and quite misleading because of the difficulty in evaluating it. Because of this risk, the judge in weighing possible prejudice against probative force under Rule 403 of the present rules exercises more control over experts than over lay witnesses."[73] As suggested by that passage, scientific evidence is often cited for its potential to mislead the jury because it may "assume a posture of mystic infallibility in the eyes of a jury of laymen."[74] Furthermore, expert testimony using such terms as "match" can be misleading unless explained.

## CABL Evidence in the Courts

Although CABL evidence has been admitted in evidence for 30 years, there are relatively few published cases on the technique. The overwhelming majority of them are homicide prosecutions,[75] some of which are capital cases. Because there are few federal homicide statutes, CABL evidence is most commonly used in state prosecutions. The courts that have addressed the admissibility of CABL evidence have admitted it—at least in the published cases.[76] CABL evidence is often used in cases in which numerous other items of evidence are introduced, but courts have sometimes indicated that it played an important role in securing a conviction.[77]

The published cases reveal a wide variety of interpretive conclusions with respect to CABL evidence. In many cases, the experts apparently have not, in their testimony, recognized the limitations of such evidence. We first describe some of the testimony and then turn to a description of permissible conclusions.

---

[73] *Daubert*, 509 U.S. at 595 (quoting Weinstein, *Rule 702 of the Federal Rules of Evidence is Sound; It Should Not Be Amended*, 138 F.R.D. 631, 632 (1991)).

[74] United States v. Addison, 498 F.2d 741, 744 (D.C. Cir. 1974). *See also* People v. King, 72 Cal. Rptr. 478, 493 (Ct. App. 1968) ("Jurors must not be misled by an 'aura of certainty which often envelops a new scientific process, obscuring its currently experimental nature.' ").

[75] *But see* United States v. Davis, 103 F.3d 660 (8th Cir. 1996) (federal trial for armed bank robbery and using a firearm during a crime of violence).

[76] As the committee was completing its report, a federal district court excluded CABL evidence under the *Daubert* standard. United States v. Mikos, 2003 WL 22922197, No. 02 CR 137 (N.D. Ill. Dec. 9, 2003).

[77] *See* Earhart v. Johnson, 132 F.3d 1062, 1068 (5th Cir. 1998) (federal habeas review) ("Given the significant role the bullet evidence played in the prosecution's case, we shall therefore assume Earhart could have made a sufficient threshold showing that he was entitled to a defense expert under Texas law."); State v. Noel, 697 A.2d 157, 160 (N.J. Super. App. Div. 1997) ("Before we address the expert-testimony problems, we note that without that testimony, the State's proofs consisted entirely of the two eyewitness identifications and defendant's possession of nine-millimeter Speers bullets. . . . Thus, with respect to the eyewitnesses, both of whom were found in the house where a suspect was believed to be and both of whom were evidently involved with drugs, one recanted and the testimony of the other was contradicted by an apparently disinterested witness."), *rev'd*, 723 A.2d 602 (N.J. 1999).

In some cases, experts have testified only that two exhibits are "analytically indistinguishable,"[78] but it is often unclear whether that was the only conclusion rendered at trial. In other cases, experts concluded that samples *could have* come from the same "source" or "batch";[79] in still others, experts stated that the samples *came* from the same source.[80]

The testimony in a number of cases goes further and refers to a "box" of ammunition (usually 50 loaded cartridges, sometimes 20). For example, two specimens

- Could have come from the same box,[81]
- Could have come from the same box or a box manufactured on the same day,[82]
- Were consistent with their having come from the same box of ammunition,[83]

---

[78]*See* Wilkerson v. State, 776 A.2d 685, 689 (Md. 2001) (The expert "concluded that all six items contained similar lead material and were probably manufactured by Remington Peters. The lead material in one bullet and one projectile was analytically indistinguishable, as was the lead in one bullet and the other two projectiles.").

[79]*See* State v. Krummacher, 523 P.2d 1009, 1012-13 (Or. 1974) (The "analyses showed that the bullet could have come from the same batch of metal as the group of bullets which was taken from defendant's home but not from the same batch as any of the other groups.").

[80]*See* United States v. Davis, 103 F.3d 660, 673-74 (8th Cir. 1996) ("He also concluded that these bullets must have been manufactured at the same Remington factory, must have come from the same batch of lead, must have been packaged on or about the same day, and could have come from the same box."); People v. Lane, 628 N.E.2d 682, 689-90 (Ill. App. 1993) ("He testified that the two bullets were analytically indistinguishable. Special Agent Riley opined that the two bullets came from the same source and that the match was *as good as he had ever seen in his twenty years with the FBI.*") (emphasis added).

[81]*See* State v. Strain, 885 P.2d 810, 817 (Utah App. 1994) ("Riley concluded that one of the bullets taken from the victim's body and the bullet taken from the gun Strain possessed when he was arrested could have come from the same box of ammunition."); State v. Jones, 425 N.E.2d 128, 131 (Ind. 1981) ("Agent Riley stated that the bullet from the victim could have come from the same box of ammunition as did the two cartridges that had bullets that matched.").

[82]*See* State v. Grube, 883 P.2d 1069, 1078 (Idaho 1994) ("He further opined that the shot shells from which the crime scene pellets came could have come from the same box as the shot shells from Grube; or were from boxes manufactured at the same place on or about the same date."); People v. Johnson, 499 N.E.2d 1355, 1366 (Ill. 1986) ("samples 'would commonly be expected to be found among bullets within the same box of cartridges with compositions just like these, and that [that is, another box of cartridges close in composition] could best be found from the same type and manufacture [sic] packaged on the same day.'"); State v. Earhart, 823 S.W.2d 607, 614 (Crim. App. Tex. 1991) ("He later modified that statement to acknowledge that analytically indistinguishable bullets which do not come from the same box most likely would have been manufactured at the same place on or about the same day; that is, in the same batch.").

[83]*See* State v. Reynolds, 297 S.E.2d 532, 534 (N.C. 1982) ("Further, neutron activation analysis revealed that the bullets taken from Morgan and Stone and the ammunition found with defendant were of the same chemical composition, consistent with their having come from the same box of ammunition.").

- Probably came from the same box,[84]
- Must have come from the same box or from another box that would have been made by the same company on the same day.[85]

The transcript in *State v. Earhart* contains the following testimony: "We can—from my 21 years experience of doing bullet lead analysis and doing research on boxes of ammunition down though the years I can determine if bullets came from the same box of ammunition. . . ."[86] In *People v. Kennedy*, the examiner testified: "If you are comparing two and they have exactly the same composition that's what you do, expect they came out of the same box."[87]

Several other (and different) statements appear in the published cases. An early case reported that the specimens "had come from the same batch of ammu-

---

[84]*See* Bryan v. Oklahoma, 935 P.2d 338, 360 (Okla. Crim. App. 1997) (FBI agent Peele testified "that the bullets from the victim, the Lincoln, the rifle, and Bryan's room all came from the same source, were manufactured in the same batch, and probably came in the same box.").

[85]*See* United States v. Davis, 103 F.3d 660, 666-67 (8th Cir. 1996) ("An expert testified that such a finding is rare and that the bullets must have come from the same box or from another box that would have been made by the same company on the same day."; the court wrote that "expert testimony demonstrated a high probability that the bullets spent at the first robbery and the last robbery originated from the same box of cartridges."); Commonwealth v. Daye, 587 N.E.2d 194, 207 (Mass. 1992) (Agent Riley testified that "two bullet fragments found in Patricia Paglia's body came from the same box of ammunition or from different boxes that were manufactured at the same place on or about the same date as a bullet retrieved from the basement of the Rye house. Riley further testified that three other bullets found in Patricia Paglia's body 'could have come from the same box of ammunition' as the two bullet fragments mentioned above."); State v. King, 546 S.E.2d 575, 584 (N.C. 2001) (Kathleen Lundy "opined that, based on her lead analysis, the bullets she examined either came from the same box of cartridges or came from different boxes of the same caliber, manufactured at the same time.").

[86] Testimony of John Riley, State v. Earhart, No. 4064, Dist Ct. Lee County, 21st Judicial Dist., Texas, Transcript at 5248-49; State v. Earhart, 823 S.W.2d 607 (Crim. App. Tex. 1991). See also Transcript at 5258 ("Well, bullets that are—that have analytically indistinguishable compositions or compositions that are generally similar typically are found within the same box of ammunition and that is the case that we have here. Now, bullets that are the same composition can also be found in other boxes of ammunition, but it's most likely those boxes would have been manufactured at the same place on or about the same date."); Testimony of John Riley, State v. Mordenti, Florida: "It's my opinion that all of those bullets came from the same box of ammunition. Now, I have to put one condition on that. And that is if they didn't come from the same box of ammunition . . . then they came from another box that was manufactured at the same place on or about the same date. And the reason I have to say that is when these cartridges were manufactured at Remington Peters, they obviously loaded more boxes than one that had this composition of bullets in it." Transcript at 480.

But see testimony of Charles Peters, Commonwealth v. Wilcox, Kentucky, Feb. 28, 2002 (Daubert hearing: "We have never testified, to my knowledge, that that bullet came from that box. We'd never say that. All we are testifying is that that bullet, or that victim fragment or something, the bullet, either came from that box or the many boxes that were produced at the same time." Transcript at 1-2.)

[87]Testimony of Ernest Peele, People v. Kennedy, No. 95CR4541, Dist. Ct., El Paso County, Colorado, July 31, 1997, Transcript.

nition: they had been made by the same manufacturer on the same day and at the *same hour*."[88] One case reports the expert's conclusion with a statistic.[89] In another case, the expert used the expressions "rare finding"[90] and "a very rare finding".[91] In still another case, the expert "opined that the same company produced the bullets at the same time, using the same lead source. Based upon Department of Justice records, she opined that an overseas company called PMC produced the bullets around 1982."[92]

In recent years, testimony appears to have become more limited. A 2002 FBI publication states the conclusion as follows: "Therefore, they *likely* originated from the same manufacturer's source (melt) of lead."[93] Testimony to the same effect has also been proffered.[94]

Recent laboratory reports reviewed by the committee contain the following conclusion: "The specimens within a composition group are analytically indistinguishable. Therefore, they originated from the same manufacturer's source (melt) of lead."[95] Another laboratory report used more cautious language: "This is consistent with the specimens within those groups originating from the same manufacturer's source (melt) of bullet lead."[96]

The most recent edition of the *FBI Handbook of Forensic Sciences* contains the following comment: "Differences in the concentrations of manufacturer-controlled elements and uncontrolled trace elements provide a means of differentiating among the lead of manufacturers, among the leads in individual manu-

---

[88]Brown v. State, 601 P.2d 221, 224 (Alaska 1979) (emphasis added) (unclear whether an FBI examiner was the expert).

[89]State v. Earhart, 823 S.W.2d 607, 614 (Crim. App. Tex. 1991) ("He concluded that the likelihood that two .22 caliber bullets came from the same batch, based on *all* the .22 bullets made in one year, is approximately .000025 percent, 'give or take a zero.' He subsequently acknowledged, however, that the numbers which he used to reach the .000025 percent statistic failed to take into account that there are different types of .22 caliber bullets made each year—.22, .22 long, and .22 long rifle. Agent Riley ultimately testified that there could be several hundred thousand bullets per batch, but with some variation in the elemental composition within the batch.").

[90]United States v. Davis, 103 F.3d 660, 666 (8th Cir. 1996) ("The bullets from the box found in the Nissan were determined to be analytically indistinguishable from the bullets recovered at the 74th Street Mid City Bank and the 42nd Street Mid City Bank. An expert testified that such a finding is rare and that the bullets must have come from the same box or from another box that would have been made by the same company on the same day.").

[91]*Id.* at 667.

[92]People v. Villarta, 2002 Cal. App. Unpub. Lexis 4776 (murder).

[93]Charles A. Peters, *The Basis for Compositional Bullet Lead Comparisons*, 4 Forensic Sci. Communications No. 3, at 5 (July 2002) (emphasis added).

[94]Testimony of Charles Peters, Commonwealth v. Wilcox, Kentucky, Feb. 28, 2002, Transcript (trial testimony): "Well, bullets that are analytically indistinguishable likely come from the same molten lead sources of lead, uh, as opposed to bullets that have different composition come from different, uh, melts of lead."

[95]State v. Anderson, Mahoning County, Ohio, March 19, 2001, Dr. Diana Grant (examiner).

[96]People v. Garner, Colorado, Dec. 11, 1998, Kathleen M. Lundy (examiner).

facturer's production lines, and among specific batches of lead in the same production line of a manufacturer."[97]

The opinions in some cases indicate that prosecutors and courts have overstated the probative impact of matching evidence. For example, in its appellate division brief in *State v. Noel*,[98] "the State asserted that this testimony is reliable scientific proof not only that the bullets 'came from the same source of lead at the manufacturer' but were 'sold in the same box.'" Part of the problem in this case was the prosecutor's summation, which made this argument. The intermediate appellate court believed that the argument was prejudicially misleading,[99] but the New Jersey Supreme Court, although conceding that the argument may have been "excessive," held that it might pass as "fair comment."[100] Similarly, in *United States v. Davis*,[101] the court wrote that "the evidence made it more probable than not that the expended bullets originated from the cartridge box found in the Nissan."[102] The committee has made several recommendations (see infra) concerning how trial testimony should be presented.

---

[97]FBI Handbook of Forensic Sciences 36 (rev. 1999). An earlier edition stated: "Analysis may determine that the composition of the bullet and or fragment is identical to the composition of the recovered ammunition. Although circumstantial, lead composition information is often useful to link a suspect to a shooting, and similar information may be determined from an analysis of shot-pellets and slugs." F.B.I. Handbook of Forensic Science 57 (rev. 1994).

[98]723 A.2d 602, 608 (N.J. 1999) (dissent).

[99]State v. Noel, 697 A.2d 157, 165 (N.J. Super. App. Div. 1997):

> Beyond the inherent problems with the expert testimony itself, we are also persuaded that the prosecutor's "snowflake or fingerprint" comment during closing must necessarily have further misled the jury in its task of assessing the probative value of Peters' identical-composition testimony. We recognize that to some extent the comment did not actually mischaracterize the testimony that the batches were most likely unique, although there was no real evidential basis for the "millions of batches" comment. The point, of course, is that the relationship of batches to billets to bullets was already confusing enough and insufficiently developed by the expert testimony. Thus, the clear import of the fingerprint and snowflake comparison was to suggest to the jury a scientific certainty in the inference that defendant had possessed both sets of bullets and to suggest to the jury a conclusiveness of that inference that clearly was not warranted. We conclude, therefore, that no matter how indulgently we might view the problems with the expert testimony itself, the prosecutor's summation, uncorrected by the court on defendant's objection, injected a high degree of prejudice into this trial.

[100]State v. Noel, 723 A.2d 602, 607 (N.J. 1999):

> In overruling defendant's objection in the prosecutor's final statement to the analogy between snowflakes and bullets, the trial court characterized the statement as a "metaphor." In his own closing argument, defense counsel, apparently anticipating the prosecutor's summation, argued that many boxes contain bullets matching the ones at issue. That argument directed the jury's attention to the issue that concerns the dissent, "whether too many bullets were in circulation to justify any conclusive inference of guilt." During the course of the trial, moreover, defense counsel vigorously cross-examined Peters. Finally, nothing prevented defense counsel from introducing evidence contradicting Peters's testimony or from requesting a charge on the jury's use of that testimony if it found the evidence to be unreliable or misleading.

[101]103 F.3d 660 (8th Cir. 1996).

[102]*Id.* at 674. The expert testified only that the bullets were analytically indistinguishable, that such a finding is rare, and "that the bullets must have come from the same box or from another box that would have been made by the same company on the same day." There may have been hundreds or thousands of other boxes manufactured that day.

## EVALUATION

CABL involves three steps: chemical analysis, statistical analysis, and the interpretation of data derived from them. As one commentator noted when evidence based on neutron activation analysis (NAA) was first introduced, "most of the legal problems surrounding NAA [now inductively coupled plasma-optical emission spectroscopy (ICP-OES)] do not involve its validity as a technique of chemical analysis. Rather, *interpretation* of the results of the chemical analysis—the relevance of the results to a particular legal issue—causes most of the difficulties."[103] Because the analytical technique (ICP-OES) has not been an issue, we deal here with the third step—relevance and interpretation.[104]

### Relevance

Evidence that crime scene bullets and loaded cartridges associated with a suspect came from the same melt is relevant under the definition of Rule 401, which is a low standard.[105] It has a "*tendency* to make the existence of any fact that is of consequence to the determination of the action [that is, the identity of the perpetrator] more probable . . . than it would be without the evidence."[106]

---

[103]Comment, *The Evidentiary Uses of Neutron Activation Analysis*, 59 Cal. L. Rev. 997, 998 (1971). Accordingly, the "qualifications of the expert as an analytical chemist do not necessarily establish his competence to interpret the legal relevance of his measurements." *Id.* at 1031.

[104]As discussed in Chapter 2, the analytical method, if properly applied, is reliable. The reliability of ICP has not been an issue in the cases or in the literature. *E.g.,* State v. Noel, 697 A.2d 157, 162 (N.J. Super. App. Div. 1997) ("To begin with, we have no doubt that ICP analysis of lead bullets is a process adequately accepted by the scientific community and producing sufficiently reliable results to warrant the admission of expert testimony regarding the test and the test results."), *rev'd on other grounds*, 723 A.2d 602 (N.J. 1999).

[105]*See* State v. Noel, 697 A.2d 157, 162 (N.J. Super. App. Div. 1997) ("Establishment of the fact that the two sets of bullets came from the same source of lead clearly enhances the probative weight that a jury would be inclined to accord to mere similarity of calibre and manufacture."), *rev'd on other grounds*, 723 A.2d 602 (N.J. 1999).

[106]Fed. R. Evid. 401 (emphasis added). *See* Margaret A. Berger, *Procedural Paradigms for Applying the Daubert Test,* 78 Minn. L. Rev. 1345, 1357 (1994) ("A match [sometimes] does have a 'tendency to make the existence of any fact that is of consequence to the determination of the action more probable . . . than it would be without the evidence.' We allow eyewitnesses to testify that the person fleeing the scene wore a yellow jacket and permit proof that a defendant owned a yellow jacket without establishing the background rate of yellow jackets in the community. Jurors understand, however, that others than the accused own yellow jackets. When experts testify about samples matching in every respect, the jurors may be oblivious to the probability concerns if no background rate is offered, or may be unduly prejudiced or confused if the probability of a match is confused with the probability of guilt, or if a background rate is offered that does not have an adequate scientific foundation.") (footnotes omitted).

The critical issues, however, are how *probative* such a finding is[107] and how that probative value is conveyed to the jury.

There are two aspects of relevance in this context: the likelihood that crime scene bullets came from the same CIVL as the defendant's bullets and the likelihood that the crime scene bullets came from the defendant.[108]

## Scientifically Supportable Conclusions (Same Melt)

A description of the probative force of evidence is given by the likelihood ratio for such evidence. The likelihood ratio for bullet lead match data is the probability that two bullets would match if they came from the same CIVL divided by the probability that they would match (coincidentally or through error) if they came from different CIVLs. If the likelihood ratio is much larger than 1, the fact of a match is strong evidence that the bullets came from the same CIVL; if not, the evidence is weak.[109]

To illustrate how this concept could be used quantitatively, assume for the sake of discussion that the probability that two bullets would match if they came from the same CIVL (the sensitivity of the test) is 0.90, and the probability of a match by coincidence or error of two bullets from different CIVLs (the false positive probability) is 1 in 500 or 0.002 The likelihood ratio[110] would then be 0.90/0.002 = 450. That can be interpreted in two ways: the probability of such a match is 450 times greater if the bullets came from the same melt than if they came from different melts, and the odds that the bullets came from the same melt are 450 times greater with the match evidence than without it (that is, there is no

---

[107]"It is probable that the jury's assessment of the strength of the link would be affected by whether defendant had a handful of similar bullets out of 1,000, or out of 10,000, or out of 100,000, or out of a million." State v. Noel, 697 A.2d 157, 163 (N.J. Super. App. Div. 1997), *rev'd on other grounds,* 723 A.2d 602 (N.J. 1999).

[108]The second issue is discussed below as "defendant as provider of bullets."

[109]*See* Richard O. Lempert, *Modeling Relevance*, 75 Mich. L. Rev. 1021, 1025-26 (1977) ("Where the likelihood ratio for an item of evidence differs from one, that evidence is *logically relevant.* This is the mathematical equivalent of the statement in *Federal Rules of Evidence* (FRE) 401 that 'relevant evidence' is 'evidence having any tendency to make the existence of any fact that is of consequence to the determination of the action more probable or less probable than it would be without the evidence. Hence, evidence is logically relevant only when the probability of finding that evidence given the truth of some hypothesis at issue in the case differs from the probability of finding the same evidence given the falsity of the hypothesis at issue. In a criminal trial, if a particular item of evidence is as likely to be found if the defendant is guilty as it is if he is innocent, the evidence is logically irrelevant on the issue of the defendant's guilt.").

[110]Here, the likelihood ration is not defined strictly as statisticians would use the term, but in a way that has been acceptable in court.

evidence either way on matching).[111] With either interpretation, the evidence in this example would strongly support the conclusion that the bullets came from the same CIVL.

However, in reality the sensitivity and the false positive rate of CABL as applied by the FBI are not available. Therefore, the interpretation can be given only in qualitative terms: the probability of a match is greater if the bullets came from the same CIVL than if they came from different CIVLs, and the odds that the bullets came from the same CIVL are greater with the matching evidence than without it. Note that the witness may not testify as to the probability or odds that the bullets came from the same CIVL but only, in the first interpretation, as to the *relative increase* in probability of a match if the bullets came from the same vs different CIVLs or, in the second interpretation, as to the *relative increase* in the odds that the bullets came from the same CIVL if they matched vs no evidence of match status.

The *admissibility* of the above-described evidence depends on whether the assumption made above, namely, that bullets from the same CIVL have a greater probability of having the same composition as bullets from different CIVLs, has sufficient scientific support to be reliable. That requires us to look at two assumptions currently made in the use of CABL: homogeneity within CIVLs (which affects the likelihood that two bullets from the same CIVL have the same composition), and homogeneity between CIVLs (which affects the likelihood that two bullets from different CIVLs have the same composition.)

---

[111]The latter formulation is an application of Bayes's theorem. In a convenient formulation, the theorem provides that:

*Posterior odds given the evidence = prior odds × likelihood ratio.*

In this context, the posterior odds given the evidence are the odds that the two bullets came from the same melt given that they are analytically indistinguishable ("match"); the prior odds are the odds that the bullets came from the same melt based on the other evidence in the case (such as evidence indicating that the bullets may have come from the defendant's supply); and the likelihood ratio is, as already defined, the probability that the bullets would match if they came from the same melt divided by the probability that they would match if they came from different melts. When FBI examiners find two bullets that match, they have a basis for testifying that the likelihood ratio is greater than 1, but they cannot properly testify as to the posterior probabilities that the bullets came from the same melt. Because they have no knowledge of the rest of the case, they have no basis for picking prior probabilities, which would be necessary for opining on posterior probabilities. Moreover, even if they had knowledge of the context of the case, testimony based on *their* prior probabilities would not necessarily be relevant or appropriate, because the jurors might have different priors, and the choice of a prior is not a matter of expertise. The most an expert can validly say is that the odds that the bullets came from the same melt are *increased* by the evidence of elemental similarity; this is true regardless of the level of prior odds. *See* State v. Spann, 617 A.2d 247 (N.J. 1993) (improper for expert to testify to posterior probabilities using her own prior).

• *Homogeneity within CIVLs.* FBI expert witnesses frequently imply or state in their testimony that if bullets came from the same melt,[112] they will always match, that is, the test has perfect sensitivity. A single study by FBI personnel tested the assumption of homogeneity of melts and found it to be reasonable (sensitivity more than 90 percent).[113] A study by critics of the assumption (Randich et al.) concludes that lead from a single melt can be inhomogeneous.[114] Possible reasons for this conclusion were discussed. However, no measure of sensitivity is given in the study, and the authors did not publish the standard deviations of their measurements, so it cannot be determined to what extent the differences found were analytically indistinguishable. Despite the debate, the existence of inhomogeneity in a melt should not seriously affect the probative value of the evidence and may, in some respects, enhance it. We discuss the reason for this below.

Even if there is considerable inhomogeneity in a melt, two bullets that come from one melt and that have the same composition must have come from a subpart of the melt that was homogeneous. Fewer bullets can be made from a subpart than from the whole melt, so the fact of inhomogeneity within a melt, if it exists, does not weaken the inferences that can be legitimately made about matching bullets. However, because the degree of inhomogeneity will in general not be known, it must be assumed, conservatively, that the number of bullets of the same composition is such as would be produced from an entire melt. The principal risk of inhomogeneity is a false negative—two bullets declared not to match when they come from the same melt. Under our system of justice, such errors are less objectionable than false positives because they would usually favor a suspect.

The committee has addressed the issue of homogeneity by defining a source not as a melt, but rather as a CIVL (compositionally indistinguishable volume of lead), which may be limited to a subpart of a melt.

• *False Positives.* False positives occur when a laboratory error or a coincidence (two CIVLs with analytically indistinguishable composition) causes two bullets to match. The rate of laboratory error is unknown because the FBI Laboratory does not have a program of testing by an external agency that has been designed to assess the proficiency of its examiners. The FBI's internal testing program does not appear to be designed to determine an error rate. If we

---

[112]In this case the term "melt" is used rather than CIVL because that is the term used by the FBI in their testimony. "Melt" will also be used on other occasions in this chapter when the original source uses the term.

[113]Robert D. Koons and Diana M. Grant, *Compositional Variation in Bullet Lead Manufacture*, 47 J. Forensic Sci. 950 (2002) (of 456 comparisons of bullets from common sources, differences were statistically and analytically significant in only 33).

[114]Erik Randich et al., *A Metallurgical Review of the Interpretation of Bullet Lead Compositional Analysis*, 127 Forensic Sci. Int'l 174 (2002).

assume the laboratory's error rate is in fact low (an assumption not currently grounded in evidence and made here only for the sake of the argument at hand), then the overwhelming contribution to the denominator of the likelihood ratio is CIVLs that are coincidentally identical in their composition.

The frequency of coincidentally identical CIVLs is unknown. Based on available data, the frequency of coincidental matches has been studied by the FBI. The data used in the FBI study have been further analyzed by the committee as described in Chapter 3. Those analyses have found some evidence supporting the assumption that the frequency of coincidental false positives is quite low. However, the FBI's study is weakened because (1) the data used by the FBI were culled by the Bureau from a larger data set consisting of a collection of bullets analyzed by the FBI over a period of 14 years, and the method of culling may have introduced statistical bias; (2) the 2-SD overlap and range overlap method used by the FBI for declaring a match do not have quantifiable error rates (although approximate error rates can be calculated as in Chapter 3); and (3) the FBI study has been neither peer-reviewed nor published.[115]

### *Daubert/Kumho* Factors

The *Daubert/Kumho* factors previously referred to provide an indication of whether proposed expert testimony is sufficiently reliable to be admissible at trial. They expressly apply to the federal courts, to the state courts in those states that have adopted *Daubert*, and are likely to be influential to some degree in those states retaining the *Frye* standard. We briefly examine below the assumptions of homogeneity and low false positive error rates from this perspective.[116]

- *Whether the theory can be and has been tested.* Both homogeneity and a low false positive rate are assumptions that can be and have been tested, as described above and in Chapter 3. As noted in those discussions, the tests of both assumptions have weaknesses. For the reasons stated above, the assumption of homogeneity within a melt is not crucial to the value of the evidence. The

---

[115]The authors of the Randich study claim in conclusory fashion that the rate of false positives is high but do not calculate a rate. If their data and assertions are accepted, the rate for their Table 3 would be about 1 in 500. The difference between the FBI rate and the Randich rate may be due in part to the fact that the Randich data are from only two manufacturers whereas the FBI data are from all manufacturers and cover a much longer period.

[116]One federal court of appeals has admitted CABL evidence under the *Daubert* test. United States v. Davis, 103 F.3d 660 (8th Cir. 1996). However, the court did not have the information that the committee had available to it. Moreover, the court overstated the probative value of the evidence. The court wrote: "The evidence made it more probable than not that the expended bullets originated from the cartridge box found in the Nissan." *Id.* at 674. As the committee was completing its report, a federal district court excluded CABL evidence under the *Daubert* standard. United States v. Mikos, 2003 WL 22922197, No. 02 CR 137 (N.D. Ill. Dec. 9, 2003).

assumption of a low false positive rate is important. As the analysis in Chapter 3 indicates, the statistical method used by the FBI may be leading to a false positive rate much higher than that assumed by examiners. A statistical method can be chosen to minimize the false positive rate, but this is always done at the expense of a higher false negative rate. Additional testing would be needed to fully satisfy the *Daubert/Kumho* testing requirement.

- *Whether the theory has been subjected to peer review and publication.* There are very few peer-reviewed articles on homogeneity and the rate of false positive matches in bullet lead composition.[117] Early articles focused on NAA[118] and other techniques,[119] used fewer elements in the analysis, and did not address the question of statistical interpretation. Moreover, some of the published articles appeared in FBI publications.[120] Outside reviews have only recently been published.[121] Because this evidence is less than conclusive and the case volume that utilizes this technique is low, the subject has not received the broad review that DNA testing and some other techniques have. Again, more such work would be needed to provide a strong basis for this admissibility factor.

- *Whether the theory has a known error rate.* The false positive probability due to coincidence has been estimated by the FBI, as noted above, but has not been published. Furthermore, as discussed in Chapter 3, this estimate is not

[117]Like many forensic techniques, CABL evidence gained admissibility before the demanding standards of *Daubert* were operative. The FBI has attempted to satisfy these standards through its recent publications and by referring the issue to this committee.

[118]*E.g.*, Vincent P. Guinn, *NAA of Bullet-Lead Evidence Specimens in Criminal Cases*, 72 J. Radioanal. Chem. 645 (1982); Vincent Guinn & M.A. Purcell, *A Very Rapid Instrumental Neutron Activation Analysis Method for the Forensic Comparison of Bullet-Lead Specimens*, 39 J. Radioanal. Chem. 85 (1977); A. Brandon & G. F. Piancone, *Characterization of Firearms and Bullets by Instrumental Neutron Activation Analysis*, 35 Int'l J. App. Radiat. Isot. 359 (1984).

[119]*See* M.A. Haney & J.F. Gallagher, *Differentiation of Bullets by Spark Source Mass Spectrometry*, 20 J. Forensic Sci. 484 (1975); R.L. Brunelle, C.M. Hoffman & K.B. Snow, *Comparison of Elemental Compositions of Pistol Bullets by Atomic Absorption: Preliminary Study*, 53 J. A.O.A.C. 470 (1970).

[120]*See* C.A. Peters, D.G. Havekost, & R.D. Koons, *Multi-Element Analysis of Bullet Lead by Inductively Coupled Plasma-Atomic Emission Spectrometry*, 15 Crime Laboratory Digest 33 (1988); E.R. Peele et al., *Comparison of Bullets Using the Elemental Compositions of the Lead Component*, Proc. Int'l Sym. On the Forensic Aspects of Trace Evidence, Quantico, Va., 1991; Charles A. Peters, *The Basis for Compositional Bullet Lead Comparisons*, 4 Forensic Sci. Communications No. 3 (July 2002).

[121]*See* Raymond O. Keto, *Analysis and Comparisons of Bullet Leads by Inductively-Coupled Plasma Mass Spectrometry*, 44 J. Forensic Sci. 1020, 1026 (1999) ("This data suggests [sic] that when two element signatures match, it is unlikely that the bullets originated from different sources. The extent of each particular source (i.e., the number of identical boxes by each manufacturer) and the bullets available in a particular geographic area at a particular time are all unknown factors."); Erik Randich et al., *A Metallurgical Review of the Interpretation of Bullet Lead Compositional Analysis*, 127 Forensic Sci. Int'l 174 (2002); William A. Tobin & Wayne Duerfeldt, *How Probative Is Comparative Bullet Lead Analysis*, 17 Crim. Justice 26 (Fall. 2002).

based upon an appropriately random sample of the bullet population. Laboratory error is another important factor in the false positive probability; the FBI has not estimated this factor and assumes it is essentially zero. In sum, the *Daubert/Kumho* factor requiring a theory to have a known error rate is only partially satisfied.

- *The existence and maintenance of standards controlling the technique's operation.* The FBI has standards controlling the training of examiners, the laboratory protocol, and the statistical method for declaring a match. However, the laboratory protocol needs to be revised to reflect current practice.[122] Moreover, the FBI does not have detailed standards governing the content of laboratory reports and the testimony that may be given by examiners. As a result, this *Daubert/Kumho* factor in significant part is not satisfied.
- *General acceptance in the relevant scientific or technical community.* The analytical technique used (that is, previously NAA and now ICP-OES) has general acceptance of the scientific community for this sample type. However, to the committee's knowledge the FBI is the only laboratory performing this type of lead analysis for forensic use, so any inquiry into "general acceptance" will not provide the broad consensus that this factor assumes. The fact that courts have generally admitted this testimony is not the equivalent of scientific acceptance, owing to the paucity of published data, the lack of independent research, and the fact that defense lawyers have generally not challenged the technique.[123]

The fact that the specifically mentioned *Daubert* factors are not fully satisfied does not mean that CABL evidence should not be admitted under the reliability standards of Rule 702. In *Kumho Tire*, the Court concluded "that a trial court *may* consider one or more of the more specific factors that *Daubert* mentioned when doing so will help determine that testimony's reliability. But as the Court stated in *Daubert*, the test of reliability is "flexible," and *Daubert*'s list of specific factors neither necessarily nor exclusively applies to all experts or in every case. Rather the law grants a district court the same broad latitude when it decides *how* to determine reliability as it enjoys in respect to its ultimate reliability determination."[124] However, the reliability and acceptance of the evidence would be strengthened if the FBI took the steps that the committee recommends.

---

[122]Conversations with FBI examiners indicate that crime bullets are compared one-to-one with the suspect's bullets and not with compositional groups of the suspect's bullets as specified by the present protocol.

[123]Attorneys have probably not challenged the evidence because the identifying link it provides to the same source is far from conclusive evidence that the defendant supplied the crime bullet. They often focus on the large number of bullets from a single melt rather than the technical intricacies of the matching process.

[124]*Kumho Tire,* 526 U.S. at 141-142 (emphasis in original).

### Defendant as Provider of Bullets

As noted earlier, relevance in this context depends not only on an association between the crime scene bullet and the same melt as the suspect's bullet but also on the further inference that this association suggests that the crime scene bullet came from the defendant. A conclusion that two bullets came from the same melt does not justify an expert in further testifying that this fact increases the odds that the crime bullet came from the defendant. The large number of bullets made from a single melt and the absence of information on the geographic distribution of such bullets[125] precludes such testimony *as a matter of expertise.*[126] Such an inference is a matter for the jury. An expert with distributional information might be able to provide such testimony to aid the jury.

The available data do not permit any definitive statement concerning the date of manufacture or the identity of the manufacturer based on elemental composition *alone*. However, in some cases, boxes with lot numbers are recovered, which may provide some information on this issue.[127] In other cases, physical (as opposed to chemical) characteristics of crime bullets are observed, which

---

[125]*See* Jones v. State, 425 N.E.2d 128, 135 (Ind. 1981) (dissent) ("all retailers in a particular geographic area might consequently market bullets of similar composition"); State v. Noel, 697 A.2d 157, 163 (N.J. Super. App. Div. 1997) ("[T]he enhancement value to be placed on the same-batch conclusion must be basically a statistical probability exercise, that is, an assessment by the trier of fact of how much more likely it is that both sets of bullets were defendant's because they not only matched in calibre and manufacture but also in composition. That assessment must necessarily depend on how many nine-millimeter bullets could have been produced from a single batch, what the likelihood is that those same bullets wound up for sale in the same geographical area, and what percentage of nine-millimeter bullets marketed in the Newark area came from Speers. Obviously, the strength of the link created by identical composition is a factor of how many bullets of identical composition were simultaneously available for sale in the Newark area, and, just as obviously, the statistical probability of defendant having possessed both sets of bullets declines as the number of identical bullets increases."), *rev'd on other grounds,* 723 A.2d 602 (N.J. 1999).

[126]The absence of distributional information also makes it inappropriate for an expert to testify that the probability that two bullets came from the same source if the defendant did not fire the crime bullet was described by the number of bullets made from the source divided by the total number of bullets of that type made in some period, such as 1 year.

[127]State v. Freeman, 531 N.W.2d 190, 195 & n. 5 (Minn. 1995) ("This box of 50 cartridges contained the same loading code, 2TB90L, as the empty cartridge box found in the snowbank at the scene of Freeman's arrest. This loading code indicated that the cartridges contained in both boxes were manufactured on February 9, 1982, during the second shift at Winchester's plant located in East Alton, Illinois."; "Also, both boxes were labelled with a Target price tag indicating a cost of $1.39."). Lot numbers indicate the date of packaging, not the date the bullet was produced or the date the loaded cartridge was assembled.

may augment the probative value of the evidence.[128] Also, "matches" of multiple crime scene bullets to multiple suspect's bullets from different CIVLs may add to the probative value of the evidence in a particular case.[129] Similarly, a case with a "closed set" of suspects presents a different situation.[130]

## PRETRIAL DISCOVERY

The need for pretrial disclosure of the nature and content of expert testimony is critical if the adversary system of trial is going to work. The American Bar Association (ABA) Standards note that the "need for full and fair disclosure is especially apparent with respect to scientific proof and the testimony of experts. This sort of evidence is practically impossible for the adversary to test or rebut at trial without an advance opportunity to examine it closely."[131] Never-

---

[128]Physical characteristics include, for example, the caliber of the bullet, the number of lands and grooves as well as their direction of twist, and whether the bullet was jacketed or not. In some cases, empty cartridge cases are found at crime scenes, which would reveal the caliber and manufacturer as well as other information. *E.g.,* State v. Ware, 338 N.W.2d 707, 712 (Iowa 1983) ("wadcutter bullet removed from Tappa's body"); State v. King, 546 S.E.2d 575, 583-84 (N.C. 2001) (Firearms examiner, who also testified in case, "determined that a spent round submitted to him, as well as the live rounds recovered during the investigation, were .22-caliber long-rifle bullets. According to Agent Wilkes, the live rounds he examined were similar in physical characteristics to the lead bullet projectile removed from the victim's wrist."); State v. Noel, 697 A.2d 157, 160 (N.J. Super. App. Div. 1997) ("A bag containing eighteen bullets was found in [defendant's] locker. Nine of the bullets were nine-millimeter bullets stamped with the manufacturer's name, Speers. The police had also recovered spent bullets and bullet casings at the crime scene. The shell casings were also stamped with the same manufacturer's name."), *rev'd on other grounds*, 723 A.2d 602 (N.J. 1999); State v. Krummacher, 523 P.2d 1009, 1012 (Or. 1974) ("The bullet found in Dorothy's body was identified as being a .38 caliber lubaloy copper-washed Smith and Wesson type bullet manufactured by the Western Company, which went out of business three years prior to the crimes in question."). The combination of physical characteristics and analytic indistinguishability can be powerful evidence in a particular case.

[129]Charles A. Peters, *The Basis for Compositional Bullet Lead Comparisons,* 4 Forensic Sci. Communications No. 3, at 5 (July 2002) ("Another factor that must be considered is a case where multiple shots of various calibers, manufacturers, and compositions are fired at a crime scene. If multiple compositions present in the crime-scene lead are analytically indistinguishable from lead groups in partial boxes of ammunition, it is much more likely that the crime-scene bullets came from those boxes than it is when only one compositional group is present.").

[130]A "closed set" case is one in which the universe of suspects is limited—for example, only one of two persons could have fired the crime bullet, so differentiation between ammunition from them is the principal concern.

[131]Commentary, ABA Standards Relating to Discovery and Procedure Before Trial 66 (Approved Draft 1970). *See also* Paul C. Giannelli, *Criminal Discovery, Scientific Evidence, and DNA*, 44 Vanderbilt L. Rev. 791 (1991).

theless, pretrial discovery is often less extensive in criminal litigation than in civil cases.[132]

Federal Criminal Rule 16 governs discovery in federal trials. Four distinct provisions are relevant to expert testimony: scientific reports, summaries of experts' expected testimony, other documents,[133] and independent testing.[134]

- *Reports.* Rule 16(a)(1)(F) makes the "results or reports of physical or mental examinations, and of scientific tests or experiments" discoverable. Under this provision, reports are discoverable if they are either material to the preparation of the defense or are intended for use by the prosecution as evidence in its case-in-chief at trial.[135] Unfortunately, the rule does not specify the content of a laboratory report. While the measurement data (means and standard deviations) on CABL evidence are discoverable, it is more logical and of greater use to include these data in the laboratory report.

---

[132]Opponents of liberal discovery argue that criminal discovery will encourage perjury, lead to the intimidation of witnesses, and, because of the Fifth Amendment, be a one-way street. 2 C. Wright, Federal Practice and Procedure § 252, at 36-37 (2d ed. 1982). In the case of scientific evidence, however, these arguments against criminal discovery lose whatever force they might otherwise have. The first argument fails because "it is virtually impossible for evidence or information of this kind to be distorted or misused because of its advance disclosure." Commentary, ABA Standards Relating to Discovery, supra, at 67. Moreover, it is extremely unlikely that an FBI expert will be subject to intimidation. *See also* 2 Wayne LaFave & Jerold H. Israel, Criminal Procedure § 19.3, at 490 (1984) ("Once the report is prepared, the scientific expert's position is not readily influenced, and therefore disclosure presents little danger of prompting perjury or intimidation"). Finally, the self-incrimination clause presents little impediment to reciprocal prosecution discovery of scientific proof. *See* Williams v. Florida, 399 U.S. 78 (1970). In any event, it seems unlikely that defense experts will be retesting this type of evidence.

[133]Rule 16(1)(a)(E) (formerly 16(1)(a)(C)) makes documents in the government's possession discoverable—such as bench notes and graphs that may not be part of the final report. *See* United States v. Armstrong, 517 U.S. 456, 463 (1996) ("Rule 16(a)(1)(C) authorizes defendants to examine Government documents material to the preparation of their defense against the Government's case-in-chief"); United States v. Zanfordianno, 833 F. Supp. 429, 432 (S.D.N.Y. 1993) ("A narrow view of Rule 16(a)(1)(C) is inappropriate; failure to provide reasonably available material that might be helpful to the defense and which does not pose any risks to witnesses or to ongoing investigation is contrary to requirements of due process and to the purposes of the Confrontation Clause. If an expert is testifying based in part on undisclosed sources of information, cross-examination vouchsafed by that Clause would be unduly restricted.").

[134]Independent testing has apparently not been a major issue in this context.

[135]Virtually all jurisdictions provide for the disclosure of scientific reports in the possession of the prosecution. Scientific reports also are discoverable under the ABA Standards and the Uniform Rules. ABA Standards for Criminal Justice 11-2.1(a)(iv) (3d ed. 1996) ("Any reports or statements made by experts in connection with the case, including results of physical or mental examinations and of scientific tests, experiments, or comparisons"); Unif. R. Crim. P. 421(a) (Approved Draft 1974) ("expert reports"). *See also* National Advisory Commission on Criminal Justice Standards and Goals, Courts, Standard 4.9(3) (1973).

The conclusions in laboratory reports should be expanded to include the limitations of CABL evidence.[136] In particular, a further explanatory comment should accompany the laboratory conclusions to portray the limitations of the evidence. Moreover, a section of the laboratory report translating the technical conclusions into language that a jury could understand would greatly facilitate the proper use of this evidence in the criminal justice system.[137] Finally, measurement data (means and standard deviations) for all of the crime scene bullets and those deemed to match should be included.

- *Summaries.* Rule 16(a)(1)(G) requires the government, on defense request, to disclose a written summary of the testimony of the experts that it intends to use during its case-in-chief. The summary must describe the witnesses' opinions, the bases of and reasons for the opinions, and the witnesses' qualifications. This provision was intended to "expand federal criminal discovery" in order to "minimize surprise that often results from unexpected expert testimony, reduce the need for continuances, and to provide the opponent with a fair opportunity to test the merit of the expert's testimony through focused cross-examination."[138] Although the ABA Standards recommend this type of discovery,[139] most states do *not* have comparable provisions.

- *Conclusions.* Like the NRC's Committee on DNA Technology in Forensic Science, the present committee concludes that broad discovery is needed to the extent feasible: "The prosecutor has a strong responsibility to reveal fully to defense counsel and experts retained by the defendant all material that might be necessary in evaluating the evidence."[140] As one court put it,

---

[136]Professor Anna Harrison, Mount Holyoke College, during a symposium on discovery, remarked: "Then the information you are receiving *is not scientific information.* For a report from a crime laboratory to be deemed competent, I think most scientists would require it to contain a minimum of three elements: (a) a description of the analytical techniques used in the test requested by the government or other party, (b) the quantitative or qualitative results with any appropriate qualifications concerning the degree of certainty surrounding them, and (c) an explanation of any necessary presumptions or inferences that were needed to reach the conclusions." *Symposium on Science and the Rules of Legal Procedure*, 101 F.R.D. 599, 632 (1984) (emphasis added).

[137]This recommendation will reduce the potentially misleading character of the evidence. See discussion of prosecution summary in *State v. Noel*, supra.

[138]Fed. R. Crim. P. 16, advisory committee's note, *reprinted at* 147 F.R.D. at 473.

[139]ABA Standards for Criminal Justice 11-2.1(a)(iv) (3d ed. 1996) ("With respect to each expert whom the prosecution intends to call as a witness at trial, the prosecutor should also furnish to the defense a curriculum vitae and a written description of the substance of the proposed testimony of the expert, the expert's opinion, and the underlying basis of that opinion.").

[140]National Research Council, DNA Technology in Forensic Science 146 (1992). *See also id.* at 105 ("Case records—such as notes, worksheets, autoradiographs, and population databanks—and other data or records that support examiners' conclusions are prepared, retained by the laboratory, and made available for inspection on court order after review of the reasonableness of a request."). The 1996 DNA report contains the following statement on discovery: "Certainly, there are no strictly scientific justifications for withholding information in the discovery process, and in Chapter

"there are no scientific grounds for withholding information in the discovery process."[141]

A statement of the limitations of CABL evidence should be included in the laboratory report. Providing an express statement of the limitations of the technique in the laboratory report not only provides notice to the parties, it affords substantial protection for experts from overreaching by attorneys. Experts are sometimes pressured by the prosecutor to "push the envelope"—not a surprising occurrence in the adversary system.[142] ABA Criminal Justice Standard 3-3.3(a) states: "A prosecutor who engages an expert for an opinion should respect the independence of the expert and should not seek to dictate the formation of the expert's opinion on the subject. To the extent necessary, the prosecutor should explain to the expert his or her role in the trial as an impartial expert called to aid the fact finders. . . ." The commentary to this standard states: "Statements made by physicians, psychiatrists, and other experts about their experiences as witnesses in criminal cases indicate the need for circumspection on the part of prosecutors who engage experts. Nothing should be done by the prosecutor to cast suspicion on the process of justice by suggesting that the expert color an opinion to favor the interests of the prosecutor."[143]

## FINDINGS AND RECOMMENDATIONS

**Finding:** Variations among and within lead bullet manufacturers makes any modeling of the general manufacturing process unreliable and potentially misleading in CABL comparisons.

**Recommendation:** Expert witnesses should define the range of "compositionally indistinguishable volumes of lead" (CIVL) that could make up the source of analytically indistinguishable bullets, because of variability in the bullet manufacturing process.

---

3 we discussed the importance of full, written documentation of all aspects of DNA laboratory operations. Such documentation would facilitate technical review of laboratory work, both within the laboratory and by outside experts. . . . Our recommendation that all aspects of DNA testing be fully documented is most valuable when this documentation is discoverable in advance of trial." National Research Council, The Evaluation of Forensic DNA Evidence 167-69 (1996).

[141]State v. Tankersley, 956 P.2d 486, 495 (Ariz. 1998).

[142]*See* Troedel v. Wainwright, 667 F. Supp. 1456, 1459 (S.D. Fla. 1986) (gunshot residue case) ("Next, as Mr. Riley candidly admitted in his deposition, he was 'pushed' further in his analysis at Troedel's trial than at Hawkins' trial. Furthermore, at the March 26th evidentiary hearing held before this Court, one of the prosecutors testified that, at Troedel's trial, after Mr. Riley had rendered his opinion which was contained in his written report, the prosecutor *pushed* to 'see if more could have been gotten out of this witness.' When questioned why, in the Hawkins trial, he did not use Mr. Riley's opinion that Troedel had fired the weapon, the prosecutor responded he did not know why."), *aff'd*, 828 F.2d 670 (11th Cir. 1987).

[143]Commentary, ABA Criminal Justice Standard 3-3.3(a) at 59.

**Finding:** The committee's review of the literature and discussions with manufacturers indicates that the size of a CIVL ranges from 70 lbs in a billet to 200,000 lbs in a melt. That is equivalent to 12,000 to 35 million 40-grain, .22 caliber longrifle bullets from a CIVL compared with a total of 9 billion bullets produced each year.

**Finding:** CABL is sufficiently reliable to support testimony that bullets from the same compositionally indistinguishable volume of lead (CIVL) are more likely to be analytically indistinguishable than bullets from different CIVLs. An examiner may also testify that having CABL evidence that two bullets are analytically indistinguishable increases the probability that two bullets came from the same CIVL, versus no evidence of match status.

**Recommendation:** Interpretation and testimony of examiners should be limited as described above and assessed regularly.

**Finding:** Although it has been demonstrated that there are a large number of different compositionally indistinguishable volumes of lead (CIVLs), there is evidence that bullets from different CIVLs can sometimes coincidentally be analytically indistinguishable.

**Recommendation:** The possible existence of coincidentally indistinguishable CIVLs should be acknowledged in the laboratory report and by the expert witness on direct examination.

**Finding:** The available data do not support any statement that a crime bullet came from, or is likely to have come from, a particular box of ammunition, and references to "boxes" of ammunition in any form is seriously misleading under Federal Rule of Evidence 403.[144] Testimony that the crime bullet came from the defendant's box or from a box manufactured at the same time is also objectionable because it may be understood as implying a substantial probability that the bullet came from defendant's box.

**Finding:** Compositional analysis of bullet lead data alone do not permit any definitive statement concerning the date of bullet manufacture.

---

[144]Testimony of Vincent Guinn, United States v. Jenkins, CR. No. 3:96-358, U.S. Dist. Ct., South Carolina, Columbia Div., Sept. 30, 1997, Transcript at 151 (Question: "Can you conclude if they match that the two bullets came from the same box of lead? [Answer:] "No, you can never do that. Every time they make a run from one particular melt, we are talking about a ton or more of lead involved. You can make an awful lot of bullets out of a ton of lead. So they get put in all these boxes and so on. . . . So, well, typically, for example, a one ton melt of lead will produce enough bullets, if it were just used itself, make enough bullets to fill something like 2,000 boxes of 50.").

**Finding:** Detailed patterns of distribution of ammunition are unknown, and as a result an expert should not testify as to the probability that a crime scene bullet came from the defendant.[145] Geographic distribution data on bullets and ammunition are needed before such testimony can be given.

**Recommendation:** The conclusions in laboratory reports should be expanded to include the limitations of compositional analysis of bullet lead evidence.[146] In particular, a further explanatory comment should accompany the laboratory conclusions to portray the limitations of the evidence. Moreover, a section of the laboratory report translating the technical conclusions into language that a jury could understand would greatly facilitate the proper use of this evidence in the criminal justice system.[147] Finally, measurement data (means and standard deviations) for all of the crime scene bullets and those deemed to match should be included.

---

[145]*See* State v. Noel, 697 A.2d 157, 162 (N.J. Super. App. Div. 1997) ("Nor was any testimony offered as to marketing, that is, whether, as seems likely, bullets from the same billets would be shipped together by the manufacturer and hence that there would be a concentration of such bullets in a specific geographical region."), *rev'd on other grounds*, State v. Noel, 723 A.2d 602 (N.J. 1999).

The defense attorney in United States v. Jenkins, CR. No. 3:96-358, U.S. Dist. Ct., South Carolina, Columbia Div., Sept. 30, 1997, argued: "No company has still today provided us with any information from which we know whether all of this ammunition ended up in Columbia, South Carolina, or whether it was randomly distributed all over the country." Transcript at 157.

Testimony of Charles Peters, Commonwealth v. Wilcox, Kentucky, Feb. 28, 2002, Transcript, (Daubert hearing & trial testimony). Question: "And do we have any information as to the geographic distribution of these bullets?" Peters: . . . "Uh, I, I don't know the information. I, uh, obviously, uh, uh, to answer that question would bring somebody in from PMC."

[146]Professor Anna Harrison, Mount Holyoke College, during a symposium on discovery, remarked: "Then the information you are receiving *is not scientific information.* For a report from a crime laboratory to be deemed competent, I think most scientists would require it to contain a minimum of three elements: (a) a description of the analytical techniques used in the test requested by the government or other party, (b) the quantitative or qualitative results with any appropriate qualifications concerning the degree of certainty surrounding them, and (c) an explanation of any necessary presumptions or inferences that were needed to reach the conclusions." *Symposium on Science and The Rules of Legal Procedure*, 101 F.R.D. 599, 632 (1984) (emphasis added).

[147]This recommendation will reduce the potentially misleading character of the evidence. See discussion of prosecution summary in *State v. Noel*, supra.

# 5

# Major Findings and Recommendations

It is the conclusion of the committee that, in many cases, CABL is a reasonably accurate way of determining whether two bullets could have come from the same compositionally indistinguishable volume of lead. It may thus in appropriate cases provide additional evidence that ties a suspect to a crime, or in some cases evidence that tends to exonerate a suspect. CABL does not, however, have the unique specificity of techniques such as DNA typing to be used as stand-alone evidence. It is important that criminal justice professionals and juries understand the capabilities as well as the significant limitations of this forensic technique. The value and reliability of CABL will be enhanced if the recommendations set forth in this report are followed.

The major findings and recommendations made by the committee in Chapters 2 through 4 are collected here.

**Finding:** The current analytical technology used by the FBI—inductively coupled plasma-optical emission spectroscopy (ICP-OES)—is appropriate and is currently the best available technology for the application.

**Recommendation:** The FBI Laboratory's analytical protocol should be revised to contain all details of the inductively coupled plasma-optical emission spectroscopy (ICP-OES) procedure and to provide a better basis for the statistics of bullet comparison. Revisions should include:

(a) Determining and documenting the precision and accuracy of the ICP-OES method and the concentration range of all seven elements to which the method is applicable.

(b) Adding data on the correlation of older neutron activation analysis and more recent ICP-OES results and any additional data that address the accuracy or precision of the method.

(c) Writing and documenting the unwritten standard practice for the order of sample analysis.

(d) Modifying and validating the digestion procedure to assure that all of the alloying elements and impurities in all samples (soft lead and hard lead) are dissolved without loss.

(e) Using a more formal control-chart system to track trends in the procedure's variability.

(f) Defining a mechanism for validation and documentation of future changes.

**Recommendation:** Because an important source of measurement variation in quality-assurance environments may be the analyst who makes the actual measurements, measurement *repeatability* (consistency of measurements made by the same analyst) and *reproducibility* (consistency of measurements made by different analysts) need to be quantified through *Gage R & R studies.* Such studies should be conducted for Federal Bureau of Investigation (FBI) comparison procedures.

**Recommendation:** The FBI's documented analytical protocol should be applied to *all* samples and should be followed by *all* examiners for *every* case.

**Recommendation:** A formal and documented comprehensive proficiency test of each examiner needs to be developed by the FBI. This proficiency testing should ensure the ability of the analyst to distinguish bullet fragments that are compositionally indistinguishable from fragments with similar but analytically distinguishable composition. Testing could be internal or external (for example, conducted by the National Institute of Standards and Technology), and test results should be maintained and provided as appropriate. Proficiency should be tested regularly.

**Recommendation:** The FBI should publish the details of its CABL procedure and the research and data that support it in a peer-reviewed journal or at a minimum make its analytical protocol available through some other public venue.

**Recommendation:** The conclusions in laboratory reports should be expanded to include the limitations of compositional analysis of bullet lead evidence. In particular, a further explanatory comment should accompany the laboratory conclusions to readily portray the limitations of the evidence. Moreover, a section of the laboratory report translating the technical conclusions into language that a jury could understand would greatly facilitate the proper use of this evidence in the criminal justice system. Finally, measurement data (means and standard deviations) for all of the crime scene bullets and those deemed to match should be included.

**Recommendation:** The FBI should continue to measure the seven elements As, Sb, Sn, Cu, Bi, Ag, and Cd as stated in the current analytical protocol.

**Recommendation:** The FBI should evaluate the potential gain from the use of high-performance inductively coupled plasma-optical emission spectroscopy because improvement in analytical precision may provide better discrimination.

**Recommendation:** The committee recommends that the FBI estimate within-bullet standard deviations on separate elements and correlations for element pairs, when used for comparisons among bullets, through use of pooling over bullets that have been analyzed with the same ICP-OES measurement technique. The use of pooled within-bullet standard deviations and correlations is strongly preferable to the use of within-bullet standard deviations that are calculated only from the two bullets being compared. Further, estimated standard deviations should be charted regularly to ensure the stability of the measurement process; only standard deviations within control-chart limits are eligible for use in pooled estimates.

**Recommendation:** The committee recommends that the FBI use either the $T^2$ test statistic or the successive t-test statistics procedure in place of the 2-SD overlap, range overlap, and chaining procedures. The tests should use pooled standard deviations and correlations, which can be calculated from the relevant bullets that have been analyzed by the FBI Laboratory. Changes in the analytical method (protocol, instrumentation, and technique) will be reflected in the standard deviations and correlations, so it is important to monitor these statistics for trends and, if necessary, to recalculate the pooled statistics.

**Recommendation:** To confirm the accuracy of the values used to assess the measurement uncertainty (within-bullet standard deviation) in each element, the committee recommends that a detailed statistical investigation using the FBI's historical dataset of over 71,000 bullets be conducted. To confirm the relative accuracy of the committee's recommended approaches to those used by the FBI, the cases that match using the committee's recommended approaches should be compared with those obtained with the FBI approaches, and causes of discrepancies between the two approaches—such as excessively wide intervals from larger-than-expected estimates of the standard deviation, data from specific time periods, or examiners—should be identified. As the FBI adds new bullet data to its 71,000+ data set, it should note matches for future review in the data set, and the statistical procedures used to assess match status.

**Recommendation:** The FBI's statistical protocol should be properly documented and followed by *all* examiners in *every* case.

**Finding:** Variations among and within lead bullet manufacturers make any modeling of the general manufacturing process unreliable and potentially misleading in CABL comparisons.

**Finding:** CABL is sufficiently reliable to support testimony that bullets from the same compositionally indistinguishable volume of lead (CIVL) are more likely to be analytically indistinguishable than bullets from different CIVLs. An examiner may also testify that having CABL evidence that two bullets are analytically indistinguishable increases the probability that two bullets come from the same CIVL, versus no evidence of match status.

**Recommendation:** Interpretation and testimony of examiners should be limited as described above, and assessed regularly.

**Recommendation:** Expert witnesses should define the range of "compositionally indistinguishable volumes of lead" (CIVL) that could make up the source of analytically indistinguishable bullets, because of variability in the bullet manufacturing process.

**Finding:** The committee's review of the literature and discussions with manufacturers indicates that the size of a CIVL ranges from 70 lbs in a billet to 200,000 lbs in a melt. That is equivalent to 12,000 to 35 million 40-grain, .22 caliber longrifle bullets from a CIVL compared with a total of 9 billion bullets produced each year.

**Finding:** Although it has been demonstrated that there are a large number of different compositionally indistinguishable volumes of lead (CIVLs), there is evidence that bullets from different CIVLs can sometimes coincidentally be analytically indistinguishable.

**Recommendation:** The possible existence of coincidentally indistinguishable CIVLs should be acknowledged in the laboratory report and by the expert witness on direct examination.

**Finding:** Compositional analysis of bullet lead data alone does not permit any definitive statement concerning the date of bullet manufacture.

**Finding:** Detailed patterns of distribution of ammunition are unknown, and as a result, an expert should not testify as to the probability that a crime scene bullet came from the defendant. Geographic distribution data on bullets and ammunition are needed before such testimony can be given.

**Finding:** The available data do not support any statement that a crime bullet came from, or is likely to have come from, a particular box of ammunition, and references to "boxes" of ammunition in any form are seriously misleading under Federal Rule of Evidence 403. Testimony that the crime bullet came from the defendant's box or from a box manufactured at the same time, is also objectionable because it may be understood as implying a substantial probability that the bullet came from defendant's box.

# APPENDIXES

# A

# Statement of Task

A committee will be appointed to assess the validity of the scientific basis for the use of elemental composition determination to compare lead alloy-based items of evidence. The following three areas will be addressed:

- *Analytical method.* Is the method analytically sound? What are the relative merits of the methods currently available? Is the selection of elements used as comparison parameters appropriate? Can additional useful information be gained by measurement of isotopic compositions?
- *Statistics for comparison.* Are the statistical tests used to compare two samples appropriate? Can known variations in compositions introduced in manufacturing processes be used to model specimen groupings and provide improved comparison criteria?
- *Interpretation issues.* What are the appropriate statements that can be made to assist the requester in interpreting the results of compositional bullet lead comparison, both for indistinguishable and distinguishable compositions? Can significance statements be modified to include effects of such factors as the analytical technique, manufacturing process, comparison criteria, specimen history, and legal requirements?

This committee will prepare an unclassified, written report at the end of the study.

# B

# Committee Membership

**Kenneth O. MacFadden,** Chair, is an Independent Consultant in Research and Analytical Management. Prior to this he was Vice President of Advanced Materials and Devices at Honeywell, Inc. In this position MacFadden was responsible for the materials and sensors research in the Corporate Research Laboratories at Honeywell. Before taking this psition in 1997, he was Vice President, Research Division at W.R. Grace & Co., where he was responsible for Analytical Research and for new product and process development in electrochemistry, bioproducts, catalysis, and polymer products. As director of analytical research, a position he assumed in 1984, he was responsible for corporate analytical support to the research division. This support included chemical and physical characterization of organic, inorganic, and biochemical materials, and compositional analysis. Other previous positions include Manager, Industrial Chemicals Research and Manager, Analytical Services at Air Products & Chemicals Inc. In the latter unit, services provided included routine chemical and physical analysis of polymers, methods development, mass spectrometric analysis, corrosion testing, polymer characterization, and environmental methods development. He has served on the Committee of Corporation Associates of the American Chemical Society and was a member of the NRC Panel for Chemical Science and Technology from 1992 to 1997 and served as Vice Chair (1995) and Chair (1996) of that panel. He was also Chair for the NRC Panel for NIST Services in 2002. He is nominated as chair because of his background in analytical chemistry, his experience running an analytical chemistry unit, and his demonstrated success in chairing NRC activities.

**A. Welford Castleman, Jr.** (NAS), a member of the Board on Chemical Sciences and Technologies, received a B.Ch.E. from Rensselaer Polytechnic Institute in 1957 and his Ph.D. (1969) degree at the Polytechnic Institute of New York. He has been on the staff of the Brookhaven National Laboratory (1958–1975), Adjunct Professor in the Departments of Mechanics and Earth and Space Sciences, State University of New York, Stony Brook (1973–1975), and Professor of Chemistry and Fellow of CIRES, University of Colorado, Boulder (1975–1982). In 1982 he accepted a professorship in the Department of Chemistry at The Pennsylvania State University, and was given the distinction of the Evan Pugh Professor title in 1986. In 1999 Professor Castleman was appointed Eberly Distinguished Chair in Science and a joint professor in the Department of Physics. He is a member of the Materials Research Institute at Penn State and is currently on the Advisory Board of the Consortium for Nanostructured Materials (VCU). Castleman's awards and honors include election to the National Academy of Sciences (1998), Fellow of the American Academy for Arts and Sciences (1998), Fellow of the New York Academy of Sciences (1998), Fellow of the American Association for the Advancement of Science (1985) and the American Physical Society (1985), receipt of the Wilhelm Jost Memorial Lectureship Award from the German Chemical Society (2000), Fulbright Senior Scholar (1989), American Chemical Society Award for Creative Advances in Environmental Science and Technology (1987), Doktors Honoris Causa from the University of Innsbruck, Austria (1987), U.S. Senior Scientist von Humboldt Awardee (1986), Senior Fellow of the Japanese Society for the Promotion of Science (1985, 1997) and Sherman Fairchild Distinguished Scholar at Cal Tech (1977). He is currently serving on the editorial boards of a number of professional publications.

**Peter R. DeForest** is Professor of Criminalistics at the John Jay College of Criminal Justice, City University of New York where he has taught for 33 years. Prior to joining the faculty and helping to found the Forensic Science B.S., M.S., and Ph.D. Programs at John Jay and the City University of New York, he worked in several laboratories. He began his career in forensic science at the Ventura County Sheriff's Crime Laboratory, Ventura, California in 1960. He earned a Bachelor of Science Degree (1964) in Criminalistics and a Doctor of Criminology Degree in Criminalistics (1969) from the University of California at Berkeley. In addition to his university teaching and research activities, he also serves as a scientific consultant and expert witness for police departments, prosecutors' offices, municipal law departments, public defender agencies, and private attorneys in criminal and civil casework. He is the author or co-author of several book chapters, a textbook, and numerous articles in scientific journals. In addition to membership in several scientific societies, he is a member of the editorial board of the *Journal of Forensic Sciences*. For over ten years, dating from the inception of the American Board of Criminalistics (ABC), Dr. De Forest served as the chairman of ABC Examination Committee, which was responsible for designing and administering certification examinations in a range of forensic science specialties. He has presented lectures and workshops for several professional societies and in other universities and has served as Visiting Professor at the University of Strathclyde, Glasgow, Scotland. During the fall 1997 semester he served as Exchange Professor with the National Crime Faculty at the Police Staff College, Bramshill, England. Awards received include the Paul L. Kirk Award of the Criminalistics Section of the American Academy of *Forensic Sciences.*

**M. Bonner Denton** is Professor of Chemistry at the University of Arizona. His research interests include applying the latest technological advances in electronics, physics, optics, astronomy, acoustics, mechanical engineering and computer science toward developing new and improved spectroscopic instrumentation and analytical methods. His multifaceted but strongly interlocking program ranges from new frontiers of mass and plasma emission spectrometry through intelligent instrumentation. Professor Denton received a Bachelor of Science in Chemistry and a Bachelor of Arts in Psychology from Lamar University in Beaumont, Texas. He then attended the University of Illinois at Champaign-Urbana, receiving his Ph.D. in Analytical Chemistry. His awards include an Alfred P. Sloan Research Fellowship, an Outstanding Young Men of America Award, the 1989 ACS Division of Analytical Chemistry Award in Chemical Instrumentation, the 1991 Society of Applied Spectroscopy's Lester Strock Award, and the Spectroscopic Society of Pittsburgh's 1998 Spectroscopy Award. He has served on the Advisory Board of *Analytical Chemistry* and on the Editorial Advisory Board of the *Journal of Automatic Chemistry*, he was President of the Society for Applied Spectroscopy, and he has been appointed an Associate Editor for *Applied Spectroscopy.*

**Charles A. Evans, Jr.,** is a consultant, recently retired from Charles Evans & Associates. This company specialized in materials analysis using microanalytical techniques such as secondary ion mass spectrometry, Rutherford backscattering spectrometry, and Auger electron spectrometry. Before starting his own company, Evans held other positions as an analytical chemist, including that of professor of chemistry. He is a member of the American Chemical Society, the American Society of Mass Spectrometry, and the Microbeam Analytical Society. Evans earned both his B.A. (1964) and Ph.D. (1968) in chemistry at Cornell University.

**Michael O. Finkelstein** has a private practice specializing in statistical methods in law and civil litigation. He is also a Lecturer at the Columbia University Law School, where he teaches statistics for lawyers. Finkelstein has also been adjunct faculty at Harvard Law School, New York University Law School, and Yale Law School. He is Editor of *The Review of Securities and Commodities Regulation* and *The Review of Banking and Financial Services*, and has written four books including *Quantitative Methods in Law* and *Statistics for Lawyers*. Finkelstein was the Chairman of the Committee on Empirical Data in Legal Decision-making for the Association of the Bar of the City of New York (1977–83) and is a Member of the American Statistical Association. He served on the NAS Committee on Statistical Assessments as Evidence in Courts (1982). Finkelstein earned his A.B. from Harvard University and his J.D. from Harvard Law School.

**Paul C. Giannelli,** Professor of Law, has been a member of Case Western Reserve University's School of Law faculty since 1975 and has twice been named Teacher of the Year. He also taught at the Judge Advocate General's School, was a Fellow in the Forensic Medicine Program of the Armed Forces Institute of Pathology & George Washington University Forensic Science Program, and served as both prosecutor and defense counsel in the General Courts-Martial Jurisdiction in the Armed Forces. A prominent expert on scientific evidence, Professor Giannelli is a frequent lecturer throughout the country, and his work has been cited in hundreds of court opinions and legal articles, including the decisions of the U.S. Supreme Court. His publications in the area of criminal law, juvenile law, evidence, and particularly scientific evidence are extensive, including coauthoring nine books, dozens of articles, and chapters in books. He authors a "Scientific Evidence" column for *Criminal Justice* and a column on "Forensic Science" for the *Criminal Law Bulletin*. Professor Giannelli is a fellow of the American Academy of Forensic Sciences and serves as Counsel for the Rules of Evidence, Ohio Supreme Court Rules Advisory Committee.

**Robert R. Greenberg** is Supervisory Research Chemist and Leader of the Nuclear Methods Group, Analytical Chemistry Division at the National Institute of Standards and Technology (NIST). The Nuclear Methods Group is at the forefront of basic and applied research into high accuracy nuclear analytical techniques. Greenberg is heavily involved in the development of nuclear methods for chemical analysis, the evaluation of the sources of error and uncertainties for these analytical techniques, and the development of Standard Reference Materials certified for chemical content. Greenberg earned his B.S. in chemistry from Brooklyn College (City University of New York) and his Ph.D. in chemistry from the University of Maryland.

**James A. Holcombe** is the Chair of the Department of Chemistry and Biochemistry at the University of Texas, Austin. Holcombe's research interest centers on improvements in trace metal analysis and speciation at the ultratrace level. He also is interested in understanding basic processes that are taking place in complex analytical atomic spectroscopic techniques in an effort to improve and expand their capabilities. His ability to draw from different areas of chemistry to attack the problem under study characterizes his program. In particular his group focuses on two main areas: design of bimolecular-based chelators and electrothermal vaporization-inductively coupled plasma-mass spectrometry. Holcombe has been editor-in-chief of *Applied Spectroscopy* and has played active roles in both the American Chemical Society and the Society for Applied Spectroscopy. He earned a B.A. from Colorado College in 1970 and a Ph.D. from the University of Michigan in 1974.

**Karen Kafadar** is a Professor of Mathematics at the University of Colorado-Denver. Her research interests include robust methods; exploratory data analysis; spatial statistics; and applications in physical, engineering, and biomedical sciences. She has previously held positions as Assistant Professor, Department of Statistics at Oregon State University; Mathematical Statistician at the National Bureau of Standards; and Mathematical Statistician at Hewlett Packard Company, and was a fellow in the Biometry Branch, Division of Cancer Prevention and Control of the National Cancer Institute. Kafadar was Editor of *Technometrics* (1999–2001) and won the 2001 William G. Hunter Award from the Statistics Division of the American Society for Quality, "for excellence in statistics as a communicator, a consultant, an educator, an innovator, an integrator of statistics with other disciplines, and an implementer who obtains meaningful results." Kafadar is also a Fellow of the American Statistical Association. She earned both her B.S. in mathematics and her M.S. in statistics from Stanford University in 1975, and her Ph.D. in statistics from Princeton University in 1979.

**Charles J. McMahon, Jr.** (NAE) received his undergraduate degree in Metallurgical Engineering from the University of Pennsylvania in 1955 and his doctorate from MIT in Physical Metallurgy in 1963, after three years of service as a line officer in the U.S. Navy. He has been a member of the faculty of the Department of Materials Science and Engineering at the University of Pennsylvania since 1964 and has spent sabbaticals at the General Electric Research and Development Center; as an Overseas Fellow at Churchill College, Cambridge University; and as a Humboldt Awardee at the Institute of Metal Physics, University of Goettingen. His research centers on metals and alloys, specifically on interfacial fracture of structural materials. McMahon also has investigated using multimedia techniques in education.

**Steven R. Prescott** is currently Manager, Analytical Sciences Division, at Hercules Inc. He is responsible for the Corporate Analytical Division, providing support to R&D, manufacturing, technical service, and regulatory groups. The Division consists of 15 laboratories involved with chemical and biological analysis, spectroscopy, separations, and materials characterization. Prior to accepting this job in 1999, Prescott was the Section Head of the Corporate Analytical Department with BetzDearborn. The mission of the Department was to provide problem solving and analytical support to the water treatment, pulp and paper, and metals processing industries. At W.R. Grace & Co., Prescott held the positions of Construction Products Product Line Manager and Water Treatment Product Line Manager, in which he developed products for the construction and water treatment industries, respectively. He was also Manager-Analytical Research, involved with the chemical characterization of organic, inorganic and biological materials. Prescott earned his B.A. at Franklin and Marshall College in 1974 and his Ph.D. from The Pennsylvania State University in 1979.

**Clifford Spiegelman** is Professor of Statistics and Toxicology at Texas A&M University. His current research interests include calibration curves, nonparametric curve fitting, and applications of statistics, particularly to chemistry. He also serves as Adjunct Professor of Chemistry at Lamar University and previously held positions at Texas A&M University and in the Statistical Engineering Division of the National Bureau of Standards. Spiegelman serves as Section Editor for chemometrics in the Wiley *Encyclopedia of Environmetrics*, sat on the board of the *Journal of Chemometrics*, and held a number of positions within the American Statistical Association. He was awarded the ASA Section on the Environment's Distinguished Achievement Award in 1994. Spiegelman earned a B.S. in economics, math, and statistics from the State University of New York, Buffalo, in 1970; an M.S. in managerial economics from Northwestern University in 1973; and a Ph.D. in statistics and applied mathematics from Northwestern University in 1976.

**Raymond S. Voorhees** oversees the research and examination programs and activities of the Physical Evidence Section, U.S. Postal Inspection Service, which is responsible for nationwide crime scene response, including bombings. He also continues to work as a forensic analyst, performing chemical, instrumental, and microscopic examinations of physical evidence, and testifying about the results of such work. Before joining the U.S. Postal Inspection Service in 1983, he spent 13 years with the Metropolitan Police Department of Washington, D.C. He served previously on the NRC Committee on Tagging Black and Smokeless Powder. Voorhees earned his B.S. in Administration of Justice from American University and his M.A. in Forensic Science at the Antioch School of Law.

# C

# Committee Meeting Agendas

**National Academy of Sciences**
**500 5th Street, NW, Room 201**
**Washington DC**

**AGENDA**

**MONDAY, FEBRUARY 3**

| | | |
|---|---|---|
| 7:30 | **BREAKFAST** | |
| 8:00 | **CLOSED SESSION** | |
| 8:00 | **OPENING REMARKS** | |
| | **Welcome**<br>Kenneth O. MacFadden, Chair | |
| | **COMMITTEE INTRODUCTIONS** | **TAB 2** |
| | **INTRODUCTION TO THE NATIONAL ACADEMIES**<br>Dorothy Zolandz, Director, Board on Chemical Sciences and Technology | |
| 8:45 | **OPEN SESSION** | |
| 8:45 | **WELCOME AND INTRODUCTIONS**<br>Kenneth O. MacFadden, Chair | |
| 9:15 | **PROJECT DESCRIPTION AND GOALS**<br>Robert D. Koons, Research Chemist, Forensic Science Research Unit, FBI Academy | **TAB 3**<br>**TAB 4** |
| 9:45 | **INTRODUCTION TO THE BULLET MANUFACTURING PROCESS**<br>Kenneth D. Green, Director, Technical Affairs, Sporting Arms and Ammunition Manufacturers' Institute, Inc. | **TAB 5** |
| 10:30 | **BREAK** | |

| | | |
|---|---|---|
| 10:45 | **FBI's Analytical Technique and Statistics**<br>Robert D. Koons, Research Chemist, Forensic Science Research Unit, FBI Academy | **TAB 6** |
| 11:45 | **Lunch** | |
| 1:00 | **Admissibility of Scientific Evidence**<br>Judge Russell F. Canan, Superior Court of the District of Columbia<br>Scott Schools, First Assistant U.S. Attorney, District of South Carolina<br>Richard K. Gilbert, Esq., Attorney | **TAB 7** |
| 2:30 | **Perspective of an Expert Witness**<br>Wayne A. Duerfeldt, Laboratory Manager, Gopher Resource Corporation | **TAB 8** |
| 3:00 | **Break** | |
| 3:15 | **Closed Session** | |
| 3:15 | **General Discussion**<br>What do we know/what don't we know/what do we need regarding:<br>• Analytical technique<br>• Bullet manufacturing<br>• Legal interpretation<br>• Data quality and quantity<br>• Other technical & legal questions | |
| 5:00 | **Adjourn For Day** | |
| 6:00 | **Committee Dinner** | |

**Tuesday, February 4**

| | |
|---|---|
| 7:30 | **Breakfast** |
| 8:00 | **Closed Session** |
| 8:00 | **Discussion of Balance and Composition**<br>Dorothy Zolandz, Director, Board on Chemical Sciences and Technology |
| 8:45 | **Summary of Discussions** |
| 10:00 | **Break** |
| 10:15 | **Report Dissemination** |
| 10:30 | **Logistics—Teams, Assignments, and Schedules** |
| 11:00 | **Next Steps** |
| 12:00 | **Adjourn** |

**National Academy of Sciences**
**500 5th Street, NW, Room 213**
**Washington DC**

## AGENDA

**MONDAY, MARCH 3**

7:30 **BREAKFAST**

8:00 **CLOSED SESSION**

8:00 **OPENING REMARKS, INTRODUCTION OF COMMITTEE MEMBERS** **TAB 2**
Kenneth O. MacFadden, Chair

8:10 **OPEN SESSION**

8:10 **WELCOME AND INTRODUCTIONS**
Kenneth O. MacFadden, Chair

8:15 **STATISTICS OF BULLETS** **TAB 4**
Alicia Carriquiry, Associate Provost and Professor, Department of Statistics, Iowa State University

9:15 **LEAD AND LEAD REFINING** **TAB 5**
R. David Prengaman, President, RSR Technologies Corporation

10:15 **BREAK**

10:45 **PERSPECTIVE ON BULLET LEAD COMPOSITIONAL ANALYSIS** **TAB 6**
TBA

11:45 **LUNCH**

1:00 **FBI EXAMINERS: QUALITY CONTROL AND TESTIMONY** **TAB 7**
TBA

2:00 **CLOSED SESSION**

2:00 **BREAK**

2:15 **REPORT FROM THE SHOT SHOW**
Robert Greenberg
Clifford Spiegelman

2:45 **UPDATES FROM GROUPS**
Manufacturing
Analytical
Statistics
Interpretation

3:15 **GROUP WORK ON NEW INFORMATION**

4:30 **GROUPS REPORT TO COMMITTEE**

5:00 **ADJOURN FOR DAY**

6:00 **COMMITTEE DINNER**

**TUESDAY, MARCH 4**

7:30 **BREAKFAST**

8:00 **CLOSED SESSION**

8:00 **DISCUSSION OF BALANCE AND COMPOSITION**
Dorothy Zolandz, Director, Board on Chemical Sciences and Technology

8:30 **COMMITTEE DISCUSSION/GROUP DISCUSSIONS ON FINDINGS**

10:00 **BREAK**

10:15 **REPORT OUTLINE DEVELOPMENT**

11:00 **LOGISTICS—ASSIGNMENTS AND SCHEDULES**

11:30 **NEXT STEPS**

12:00 **ADJOURN**

**National Academy of Sciences**
**500 5th Street, NW, Room 213**
**Washington DC**

## AGENDA

**MONDAY, APRIL 14**

7:30 **BREAKFAST**

8:00 **CLOSED SESSION**

8:00 **OPENING REMARKS: REVIEW COMMITTEE CHARGE & WORK SESSION GOALS**
Kenneth O. MacFadden, Chair

8:15 **SUB-GROUP VISIT REPORTS**
Manufacturing—visit to bullet manufacturers
Analysis
Statistics—conference call with Robert Koons
Interpretation

9:30 **OPEN SESSION**

9:30 **BREAK**

9:45 **Question and Answer Session**
Robert Koons, FBI Laboratory

11:45 **Lunch**

1:00 **Closed Session**

1:00 **What we know/don't know exercise (Review master document)**

3:00 **Sub-group Work Session**

4:30 **Development of Writing Plan**

5:00 **Adjourn for Day**

6:00 **Committee Dinner**

**Tuesday, April 15**

7:30 **Breakfast**

8:00 **Closed Session**

8:00 **Sub-group Report Writing**

11:45 **Lunch**

12:45 **Committee Progress Reports**

1:30 **Discussion of Work Plan, Task Assignments and Next Meeting**

2:00 **Adjourn**

2:00 **Trip to FBI Laboratory**

**National Academy of Sciences**
**500 5th Street, NW, Room 205**
**Washington DC**

## AGENDA

**Monday, May 12**

7:30 **Breakfast**

8:00 **Closed Session**

8:00 **Opening Remarks: Review Committee Charge & Work Session Goals**
Kenneth O. MacFadden, Chair

8:15 **Report on Visit to FBI Laboratory**
Steve Prescott, Karen Kafadar, Michael Finkelstein, Bob Greenberg, Ken MacFadden

8:30 **REVIEW & DISCUSSION OF DRAFT REPORT**
Each team to give a brief overview of their Draft report, conclusions and recommendations: followed by an open discussion with the committee

8:30 **ANALYSIS**

9:15 **MANUFACTURING**

10:15 **BREAK**

10:30 **STATISTICS**

11:15 **INTERPRETATION**

12:00 **LUNCH**

1:00 **OPEN SESSION**

1:00 **QUESTION AND ANSWER SESSION**
Robert Koons, FBI Laboratory

3:00 **CLOSED SESSION**

3:00 **DISCUSSION AND RESOLUTION OF OPEN ISSUES**

5:00 **ADJOURN FOR DAY**

6:30 **COMMITTEE DINNER**

**TUESDAY, MAY 13**

7:30 **BREAKFAST**

8:00 **CLOSED SESSION**

8:00 **SUB-GROUP REPORT WRITING**

11:45 **LUNCH**

12:45 **COMMITTEE PROGRESS REPORTS**

1:15 **DISCUSSION OF REPORT DISSEMINATION**

1:30 **DISCUSSION OF WORK PLAN, REPORT SCHEDULE, TASK ASSIGNMENTS**

2:00 **ADJOURN**

# D

# Glossary

**Acronyms and Terminology**

| | |
|---|---|
| AAS | Atomic Absorption Spectrometry |
| Ag | Silver, a metallic element sometimes present as an impurity in bullet lead |
| Ammunition | The loaded "round" commonly consisting of a primed case, propellant (powder), and bullet |
| As | Arsenic, a semi-metallic element sometimes present as an impurity in bullet lead |
| Bi | Bismuth, a metallic element sometimes present as an impurity in bullet lead |
| Billet | A cylinder of lead, usually weighing about 70 lbs, that is used as the stock for an extrusion press to make wire for the production of lead bullets |
| Blast furnace | A large vertical furnace used to reduce lead ores to molten lead in which hot coke reduces the sinter roast through the formation of $CO_2$; the necessary heat is produced by the reaction of the coke with air forced into the furnace from below |
| Bullet | The lead-based projectile in small-arms ammunition |
| Bullet caliber | The diameter of the bullet, which may be expressed either as a fraction of an inch, e.g., .22 caliber means 0.22 inch diameter, or in millimeters |
| CABL | Compositional Analysis of Bullet Lead |
| Cartridge | A term used to refer either to the completely assembled ammunition or to the brass case that holds the primer and powder and is pressed onto the bullet |

| | |
|---|---|
| CCI | CCI, bullet manufacturer |
| Cd | Cadmium, a metallic element sometimes present as an impurity in bullet lead |
| CIVL | Compositionally Indistinguishable Volume of Lead |
| Compositional group | A set of bullets determined to be compositionally similar via use of the FBI's "chaining" technique |
| COV | Coefficient of Variation |
| CS | Crime Scene (bullet) |
| Cu | Copper, a metallic element used in jacketing high velocity ammunition and sometimes present as an impurity in bullet lead |
| Extruder | The machine that forces lead from a billet through an orifice or die to form a wire (much like squeezing toothpaste from a tube) |
| FBI | Federal Bureau of Investigation |
| FED | Federal, bullet manufacturer |
| Hog | A one ton casting of lead |
| ICP-MS | Inductively Coupled Plasma-Mass Spectrometry |
| ICP-OES | Inductively Coupled Plasma-Optical Emission Spectroscopy |
| Ingot | A 65–125 lb casting of lead; more generally, a casting that has solidified after having been poured from a vessel in the form of molten metal |
| Jacket | A metal external shell, often copper, surrounding the lead core of a bullet, frequently used for high velocity ammunition |
| LA | Laser Ablation |
| MC-ICP-MS | Multi-Collector-Inductively Coupled Plasma-Mass Spectrometry |
| Melt | A quantity of molten lead |
| Mold | The container into which molten metal is poured to allow it to solidify |
| NAA | Neutron Activation Analysis |
| Pb | Lead, a metallic element used to form bullets |
| PCA | Principal Components Analysis |
| Pig | A 65–125 lb. casting of lead |
| Pot | A vessel within which lead is melted |
| Pour | The action of transferring a molten metal from a vessel into an ingot mold, in which it will solidify |
| Primary lead smelter | A facility that transforms lead-bearing ore, normally a sulfide, into nearly pure lead by the steps of sintering, reduction, and refining |
| PS | Probable Suspect (bullet) |

| | |
|---|---|
| Reduction | The chemical process of converting the lead ore into molten lead |
| Refining | The process of removing unwanted contaminants by various treatments carried out on a bath of molten lead |
| REM | Remington, bullet manufacturer |
| RF | Radio-Frequency |
| RSD | Relative Standard Deviation |
| Sb | Antimony, an element used to harden lead for bullets. |
| SD | Standard Deviation |
| Secondary lead smelter | An organization that remelts scrap lead from various sources and carries out refining and alloying operations to produce lead ingots, pigs, billets, etc. of specified composition for further processing and/or product formation |
| Slug | A cylinder of lead that has been cut for an extruded wire and that approximates the size (length and diameter) of the finished bullet |
| Sn | Tin, a metallic element also used for hardening lead, but it is more expensive and less effective than antimony. Also a metal sometimes present as an impurity in bullet lead |
| SRM | Standard Reference Material |
| SSMS | Spark Source Mass Spectrometry |
| Suspect bullet | Unused cartridges in the possession of a suspect |
| Swage | An operation that involves rotary forging, employing rotating dies that periodically open and close, used to reduce the diameter of rods, wires, or tubes. (Often used in the firearms industry to mean pressing of a slug into a bullet.) |
| TIMS | Thermal Ionization Mass Spectrometry |
| WDXRF | Wavelength Dispersive X-Ray Fluorescence |
| WIN | Winchester, bullet manufacturer |
| Wire | A long piece of lead of the correct diameter used to produce a desired caliber bullet, formed by extrusion |

**Statistical Terminology**

| | |
|---|---|
| $K_a$ | critical value |
| $t_a$ | critical value |
| $\sigma$ | within-bullet standard deviation |

# E

# Basic Principles of Statistics[1]

All measurements are subject to error. Analytical chemical measurements often have the property that the error is proportional to the value. Denote the $i^{th}$ measurement on bullet $k$ as $X_{ik}$ (we will consider only one element in this discussion and hence drop the subscript $j$ utilized in Chapter 3). Let $\mu^*_{xk}$ denote the mean of all measurements that could ever be taken on this bullet, and let $\varepsilon^*_{ik}$ denote the error associated with this measurement. A typical model for analytical measurement error might be

$$X_{ik} = \mu^*_{xk} \cdot \varepsilon^*_{ik},\ i = 1,2,3;\ k = \text{number of CS bullets.}$$

Likewise, for a given PS bullet measurement, $Y_{ik}$, with mean $\mu^*_{yk}$ and error in measurement $\eta_{ik}$,

$$Y_{ik} = \mu^*_{yk} \cdot \eta^*_{ik},\ i = 1,2,3;\ k = \text{number of PS bullets.}$$

Notice that if we take logarithms of each equation, these equations become additive rather than multiplicative in the error term:

$$\log(X_{ik}) = \log(\mu^*_{xk}) + \log(\varepsilon^*_{ik})$$
$$\log(Y_{ik}) = \log(\mu^*_{yk}) + \log(\eta^*_{ik})$$

Models with additive rather than multiplicative error are the basis for most statistical procedures. In addition, as discussed below, the logarithmic transformation yields more normally distributed data as well as transformed measure-

[1]Note that the notation used in this Appendix differs from that used in the body of the report.

ments with constant variance. That is, an estimate of $\log(\mu_{xk})$ is the logarithm of the sample average of the three measurements on bullet $k$, and a plot of these log(averages) shows more normally distributed values than a plot of the averages alone. We denote the variances of $\mu_{xk} \equiv \log(\mu^*_{xk})$ and $\mu_{yk} \equiv \log(\mu^*_{yk})$ as $\sigma^2_x$ and $\sigma^2_y$, and the variances of the error terms $\varepsilon_{ik} \equiv \varepsilon^*_{ik}$ and $\eta_{ik} \equiv \eta^*_{ik}$ as $\sigma^2_e$ and $\sigma^2_n$, respectively. It is likely that the between-bullet variation is the same for the populations of both the CS and the PS bullets; therefore, since $\sigma^2_x$ should be the same as $\sigma^2_y$, we will denote the between-bullet variances as $\sigma^2_b$. Similarly, if the measurements on both the CS and PS bullets were taken at the same time, their errors should also have the same variances; we will denote this within-bullet variance as $\sigma^2_e$, or $\sigma^2$ when we are concentrating on just the within-bullet (measurement) variability.

Thus, for three reasons—the nature of the error in chemical measurements, the approximate normality of the distributions, and the more constant variance (that is, the variance is not a function of the magnitude of the measurement itself)—logarithmic transformation of the measurements is advisable. In what follows, we will assume that $x_i$ denotes the logarithm of the $i^{th}$ measurement on a given CS bullet and one particular element, $\mu_x$ denotes the mean of these log(measurement) values, and $\varepsilon_i$ denotes the error in this $i^{th}$ measurement. Similarly, let $y_i$ denote the logarithm of the $i^{th}$ measurement on a given PS bullet and the same element, $\mu_y$ denote the mean of these log(measurement) values, and $\eta_i$ denote the error in this $i^{th}$ measurement.

## NORMAL (GAUSSIAN) MODEL FOR MEASUREMENT ERROR

All measurements are subject to measurement error:

$$x_i = \mu_x + \varepsilon_i$$
$$y_i = \mu_y + \pi_i$$

Ideally, $\varepsilon_i$ and $\pi_i$ are small, but in all instances they are unknown from measured replicate to replicate. If the measurement technique is *unbiased*, we expect the mean of the measurement errors to be zero. Let $\sigma^2_\varepsilon$ and $\sigma^2_\pi$ denote the measurement errors' variances. Because $\mu_x$ and $\mu_y$ are assumed to be constant, and hence have variance 0, $Var(x_i) = \sigma^2_x = \sigma^2_\varepsilon$, and $Var(y_i) = \sigma^2_y = \sigma^2_\pi$. The distribution of measurement errors is *often* (not always) assumed to be normal (Gaussian). That assumption is often the basis of a convenient model for the measurements and implies that

$$P\{\mu_x - 1.96\sigma_x < x_i < \mu_x + 1.96\sigma_x\} = 0.95 \quad \text{(E.1)}$$

if $\mu_x$ and $\sigma_x$ are known (and likewise for $y_i$, using $\mu_y$ and $\sigma_y$). (The value 1.96 is often conveniently rounded to 2.) Moreover, $\bar{x} = \sum_{i=1}^{3} x_i / 3$ will also be normally distributed, also with mean $\mu_x$ but with a smaller variance, $\sigma_x^2/3, (SD = \sigma_x/\sqrt{3})$, therefore

$$P(\mu_x - 1.96\sigma_x/\sqrt{3} < \bar{x} < \mu_x + 1.96\sigma_x/\sqrt{3}) = 0.95.$$

Referring to Part (b) of the Federal Bureau of Investigation (FBI) protocol for forming "compositional groups" (see Chapter 3), its calculation of the standard deviation of the group is actually a standard deviation of averages of three measurements, or an estimate of $\sigma_x/\sqrt{3}$ in our notation, not of $\sigma_x$. In practice, however, $\mu_x$ and $\sigma_x$ are unknown, and interest centers not on an individual $x_i$ but rather on $\mu_x$, the mean of the distribution of the measured replicates. If we estimate $\mu_x$ and $\sigma_x$ using $\bar{x}$ and $s_x$ from only three replicates as in the current FBI procedure but still assume that the measurement error is normally distributed, then a 95 percent confidence interval for the true$\{\mu_x\}$ can be derived from Equation E.1 by rearranging the inequalities using the correct multiplier, not from the Gaussian distribution (that is, not 1.96 in Equation E.1) but rather from Student's *t* distribution, and the correct standard deviation $s_x/\sqrt{3}$ instead of $s_x$:

$$P\{\bar{x} - 4.303 s_x/\sqrt{3} < \mu_x < \bar{x} + 4.303 s_x/\sqrt{3}\} = 0.95$$
$$\Rightarrow P(\bar{x} - 2.484 s_x < \mu_x + 2.484 s_x) = 0.95.$$

Use of the multiplier 2 instead of 2.484 yields a confidence coefficient of 0.926, not 0.95.

## CLASSICAL HYPOTHESIS-TESTING: TWO-SAMPLE *t* STATISTIC

The present situation involves the comparison between the sample means $\bar{x}$ and $\bar{y}$ from two bullets. Classical hypothesis-testing states the null and alternative hypotheses as $H_0{:}\mu_x = \mu_y$ vs $H_1{:}\mu_x \neq \mu_y$ (reversed from our situation), and states that the two samples of observations (here, $x_1$, $x_2$, $x_3$ and $y_1$, $y_2$, $y_3$) are normally distributed as $N(\mu_x,\sigma_x^2)$, $N(\mu_y,\sigma_y^2)$ and $\sigma_x^2 = \sigma_y^2 = \sigma^2$. Under those conditions, $\bar{x}$, $\bar{y}$ and $s_p$ are highly efficient estimates of $\mu_x$, $\mu_y$, and $\sigma$, respectively, where $s_p$ is a pooled estimate of the standard deviation that is based on both samples:

$$s_p = \sqrt{[n_x - 1)s_x^2 + (n_y - 1)s_y^2] / (n_x + n_y - 2)}. \qquad \text{(E.2)}$$

Evidence in favor of $H_1{:}\mu_x \neq \mu_y$ occurs when $\bar{x}$ and $\bar{y}$ are "far apart." Formally, "far apart" is determined when the so-called two-sample *t* statistic (which, under $H_0$, has a central Student's *t* distribution on $n_x + n_y - 2 = 3 + 3 - 2 = 4$ degrees of freedom) exceeds a critical point from this Student's $t_4$ distribution. To ensure a false null hypothesis rejection probability of no more than $100\alpha\%$ where $\alpha$ is the probability of rejecting $H_0$ when it is correct (that is, claiming "different" when the means are equal), we reject $H_0$ in favor of $H_1$ if

$$pooled-two-sample-t = |\bar{x}-\bar{y}|/s_p\sqrt{1/n_x+1/n_y}] > t_{n_x+n_y-2,\alpha/2} \quad \text{(E.3)}$$

where $t_{n_x+n_y-2,\alpha/2}$ is the value beyond which only $100 \cdot \alpha/2\%$ of the Student's $t$ distribution (on $n_x + n_y - 2$ degrees of freedom) lies.

When $n_x = n_y = 2$ and $s_p = \sqrt{(s_x^2 + s_y^2)/2}$, Equation E.3 reduces to:

$$|\bar{x}-\bar{y}|/[s_p\sqrt{2/3}] > 2.776, for\ \alpha = 0.05. \quad \text{(E.4)}$$

This procedure for testing $H_0$ versus $H_1$ has the following property: among all possible tests of $H_0$ whose false rejection probability does not exceed $\alpha$, this two-sample Student's $t$ test has the maximum probability of rejecting $H_0$ when $H_1$ is true (that is, has the highest power to detect when $\mu_x$ and $\mu_y$ are unequal). If the two-sample $t$ statistic is less than this critical value (2.776 for $\alpha = 0.05$), the interpretation is that the data do not support the hypothesis of different means. A larger critical value would reject the null hypothesis ("same means") less often.

The FBI protocol effectively uses $s_x + s_y$ in the denominator instead of $s_p\sqrt{2/3}$ and uses a "critical value" of 2 instead of 2.776. Simulation suggests that the distribution of the ratio $(s_x + s_y)/s_p$ has a mean of 1.334 (10%, 25%, 75%, and 90% quantiles are 1.198, 1.288, 1.403, and 1.413, respectively). Substituting $[(s_x + s_y)/1.334]\sqrt{2/3}$ for $s_p\sqrt{2/3}$ suggests that the approximate error in rejecting $H_0$ when it is true for the FBI statistic, $|\bar{x} - \bar{y}|/(s_x + s_y)$, would also be 0.05 *if* it used a "critical point" of $2.776\sqrt{2/3}/1.334 = 1.70$. Replacing 1.334 with the quantiles 1.198, 1.288, 1.403, and 1.413 yields values of 1.892, 1.760, 1.616, and 1.604, respectively—all smaller than the FBI value of 2. The FBI value of 2 would correspond to an approximate error of 0.03. A larger critical value (smaller error) leads to fewer rejections of the null hypothesis, that is, more likely to claim "equality" and less likely to claim "different" when the means are the same.

If the null hypothesis is $H_0{:}\mu_x - \mu_y = \delta(\delta \neq 0)$, the two-sample $t$ statistic in Equation E.4 has a *noncentral t* distribution with noncentrality parameter $(\delta/\sigma)(n_x n_y)/(n_x + n_y)$, which reduces to $(\delta/\sigma)(n/2)$ when $n_x = n_y = n$. When the null hypothesis is $H_0{:}\ |\ \mu_x - \mu_y\ | \geq \delta$ vs $H_1{:}\ |\ \mu_x - \mu_y\ | < \delta$, the distribution of the pooled two-sided two-sample $t$ statistic (Equation E.4) has a noncentral $F$ distribution with 1 and $n_x + n_{y-2} = 2(n-1)$ degrees of freedom and noncentrality parameter $[(\delta/\sigma)n_x n_y/(n_x + n_y)]^2 = [(\delta/\sigma)n/2]^2$.

The use of Student's $t$ statistic is *valid* (that is, the probability of falsely rejecting $H_0$ when the means $\mu_x$ and $\mu_y$ are truly equal is $\alpha$) only when the $x$'s and $y$'s are normally distributed. The appropriate critical value (here, 2.776 for $\alpha = 0.05$ and $\delta = 0$) is different if the distributions are not normal, or if $\sigma_x \neq \sigma_y$, or if $H_0$: $| \mu_x - \mu_y | \geq \delta \neq 0$, or if $(s_x + s_y)/2$ is used instead of $s_p$ (Equation E.2), as is used currently in the FBI's statistical method. It also has the highest power (highest probability of claiming $H_1$ when in fact $\mu_x \neq \mu_y$, subject to the condition that the probability of erroneously rejecting $H_0$ is no more than $\alpha$.

The assumption "$\sigma_x = \sigma_y$" is probably reasonably valid if the measurement process is consistent from bullet sample to bullet sample: one would expect the error in measuring the concentration of a particular element for the crime scene (CS) bullet ($\sigma_x$) to be the same as that in measuring the concentration of the same element in the potential suspect (PS) bullet ($\sigma_y$). However, the normality assumption may be questionable here; as noted by (Ref. 1), average concentrations for different bullets tend to be lognormally distributed. That means that log(As average) is approximately normal as it is for all six other elements. When the measurement uncertainty is very small (say, $\sigma_x < 0.2$), the lognormal distribution differs little from the normal distribution (Ref. 2), so these assumptions will be reasonably well satisfied for precise measurement processes. Only a few of the standard deviations in the datasets were greater than 0.2 (see the section titled "Description of Data Sets" in Chapter 3).

The case of CABL differs from the classical situation primarily in the reversal of the null and alternative hypotheses of interest. That is, the null hypothesis here is $H_0$:$\mu_x \neq \mu_y$ vs $H_1$:$\mu_x = \mu_y$. We accommodate the difference by stating a specific relative difference between $\mu_x$ and $\mu_y$, $|\mu_x - \mu_y|$, and rely on the noncentral F distribution as mentioned above.

## EQUIVALENCE *t* TESTS[2]

An equivalence $t$ test is designed to handle our situation:

$H_0$: means are different.
$H_1$: means are similar.

Those hypotheses are quantified more precisely as

$H_0$:$|\mu_x - \mu_y| \geq \delta$.
$H_1$: means are $|\mu_x - \mu_y| < \delta$.

We must choose a value of $\delta$ that adequately reflects the condition that "two bullets came from the same compositionally indistinguishable volume of mate-

[2]Note that the form of this test is referred to as successive t-test statistics in Chapter 3. In that description, the setting of error rates is not prescribed.

rial (CIVL), subject to specification limits on the element given by the manufacturer." For example, if the manufacturer claims that the Sb concentrations in a given lot of material are 5% ± 0.20%, a value of $\delta = 0.20$ might be deemed reasonable. The test statistic is still the two-sample *t* as before, but now we reject $H_0$ if $\bar{x}$ and $\bar{y}$ are too *close*. As before, we ensure that the false match probability cannot exceed a particular value by choosing a critical value so that the probability of falsely rejecting $H_0$ (falsely claiming a "match") is no greater than $\alpha$ (here, we will choose $\alpha = 1/2{,}500 = 0.0004$ for example. The equivalence test has the property that, subject to false match probability $\leq \alpha = 0.0004$, the probability of *correctly* rejecting $H_0$ (that is, claiming that two bullets match when the means of the batches from which the bullets came are less than $\delta$), is maximized. The left panel of Figure E.1 shows a graph of the distribution of the difference $\bar{x} - \bar{y}$ under the null hypothesis that $\delta/\sigma = 0.25$ (that is, either $\mu_x - \mu_y = -0.25\sigma$, or $\mu_x - \mu_y = +0.25\sigma$) and $n = 100$ fragment averages in each sample, subject to false match probability $\leq 0.05$: the equivalence test in this case rejects $H_0$ when $|\bar{x} - \bar{y}| / (s_p\sqrt{2/100}) < 1$. The right panel of Figure E.1 shows the power of this test: when $\delta$ equals zero, the probability of correctly rejecting the null hypothesis ("means differ by more than 0.25") is about 0.60, whereas the probability of rejecting the null hypothesis when $\delta = 0.25$ is only 0.05 (as it should be, given the specifications of the test). Figure E.1 is based on the information given in Wellek (Ref. 3); similar figures apply for the case when $\alpha = 0.0004$, $n = 3$ measurements in each sample, and $\delta/\sigma = 1$ or 2.

## DIGRESSION: LOGNORMAL DISTRIBUTIONS

This section explains two benefits of transforming measurements via logarithms for the statistical analysis.

The standard deviations of measurements made with inductively coupled plasma-optical emission spectroscopy are generally proportional to their means; hence, one typically refers to *relative* error, or coefficient of variation, sometimes expressed as a percentage, $(s_x/\bar{x}) \times 100\%$. When the measurements are transformed first via logarithms, the standard deviation of the log(measurements) is approximately, and conveniently, equal to the coefficient of variation (COV), sometimes called relative error (RE), in the original scale. This can be seen easily through standard propagation-of-error formulas (Ref. 4, 5), which rely on a first-order Taylor series expansion for the transformation (here, the natural logarithm) about the mean in the original scale—

$$f(X) = f(\mu_x) + (X - \mu_x)f'(\mu_x) + \ldots$$
$$\Rightarrow Var[f(X)] \approx [f'(\mu_x)]^2 \sigma_x^2$$

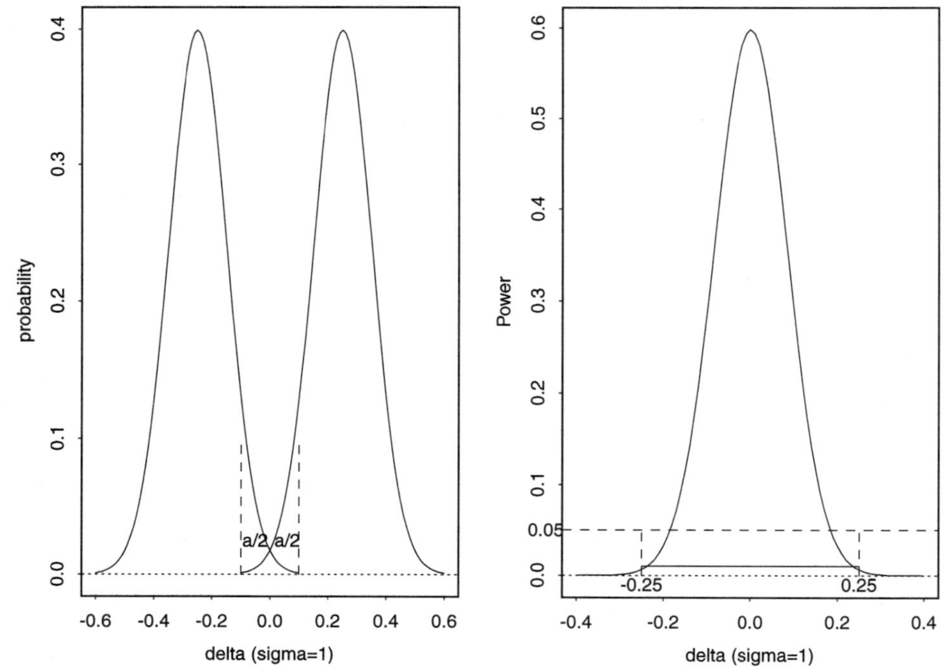


FIGURE E.1 The left panel shows a picture of the distribution of the difference $\bar{x} - \bar{y}$ under the null hypothesis that $\delta / \sigma = 0.25$ and $n = 100$ fragment averages in each sample, subject to false match probability $\leq 0.05$ : the equivalence test in this case rejects $H_0$ when $|\bar{x} - \bar{y}| / (s_p \sqrt{2/100}) < 1$. The right panel shows the power of this test: when $\delta$ equals zero, the probability of correctly rejecting the null hypothesis is about 0.60, whereas the probability of rejecting the null hypothesis when $\delta = 0.25$ is only 0.05. Figure is based on information given in Wellek (Ref. 3).

—because the variance of a constant (such as $\mu_x$) is zero. Letting $f(X) = \log(X)$, and $f'(\mu_x) = 1/\mu_x$, it follows that

$$Var[\log(X)] \approx \sigma_x^2 / \mu_x^2 \Rightarrow SD[\log(X)] \approx \sigma_x / \mu_x = COV_x = RE_x$$

Moreover, the distribution of the logarithms for each element tends to be more normal than that of the raw data. Thus, to obtain more-normally distributed data and as a by-product a simple calculation of the COV, the data should first be transformed via logarithms. Approximate confidence intervals are calculated in the log scale and then can be transformed back to the original scale via the antilogarithm, $(e^{\bar{x}-2SD}, e^{\bar{x}+2SD})$.

## DIGRESSION: ESTIMATING $\sigma^2$ WITH POOLED VARIANCES

The FBI protocol for statistical analysis estimates the variances of the triplicate measurements in each bullet with only three observations, which leads to highly variable estimates—a range of a factor of 10, 20, or even more ($\chi^2_{0.90,2}$, $\chi^2_{0.10,2}$). Assuming that the measurement variation is the same for both the PS and CS bullets, the classical two-sample $t$ statistic pools the variances into $s_p^2$ (Equation E.2), which has four degrees of freedom and is thus more stable than either individual $s_x$ or $s_y$ alone (each based on only two degrees of freedom). The pooled variance $s_p^2$ need not rely on only the six observations from the two samples if the within-replicate variance is the same for several bullets. Certainly, that condition is likely to hold if bullets are analyzed with a consistent measurement process. If three measurements are used to calculate each within-replicate standard deviation from each of, say, $B$ bullets, a better, more stable estimate of $\sigma^2$ is

$$s_p^2 = (s_1^2 + \ldots + s_B^2) / B.$$

Such an estimate of $\sigma^2$ is now based on not just 2(2) = 4 degrees of freedom, but rather $2B$ degrees of freedom. A stable and repeatable measurement process offers many estimates of $\sigma^2$ from many bullets analyzed by the laboratory over several years. The within-replicate variances may be used in the above equation. To verify the stability of the measurement process, standard deviations should be plotted in a control-chart format ($s$-chart) (Ref. 7) with limits that, if exceeded, indicate a change in precision. Standard deviations that fall within the limits should be pooled as in Equation E.3. Using pooled standard deviations guards against the possibility of claiming a match simply because the measurement variability on a particular day happened to be large by chance, creating wider intervals and hence greater chances of overlap.

To determine whether a given standard deviation, say, $s_g$, might be larger than the $s_p$ determined from measurements on $B$ previous bullets, one can compare the ratio $s_p^2/s_g^2$ with an $F$ distribution on 2 and $2B$ degrees of freedom. Assuming that the FBI has as many as 500 estimates, the 5% critical point from an $F$ distribution on two and 1,000 degrees of freedom is 3.005. Thus, if a given standard deviation is $\sqrt{3} = 1.732$ times larger than the pooled standard deviation for that element, one should consider remeasuring that element, in that the precision may be larger than expected by chance alone (5% of the time).

## REFERENCES

1. Carriquiry, A.; Daniels, M.; and Stern, H. "Statistical Treatment of Case Evidence: Analysis of Bullet Lead", *Unpublished report,* 2002.
2. Antle, C.E. "Lognormal distribution" in *Encyclopedia of Statistical Sciences, Vol 5,* Kotz, S.; Johnson, N. L.; and Read, C. B., Eds.; Wiley: New York, NY, 1985, pp. 134–136.
3. Wellek, S. *Testing Statistical Hypotheses of Equivalence* Chapman and Hall: New York, NY 2003.
4. Ku, H.H. Notes on the use of propagation of error formulas, *Journal of Research of the National Bureau of Standards-C. Engineering and Instrumentation,* 70C(4), 263–273. Reprinted in *Precision Measurement and Calibration: Selected NBS Papers on Statistical Concepts and Procedures,* NBS Special Publication 300, Vol. 1, H.H. Ku, Ed., 1969, 331–341.
5. Cameron, J.E. "Error analysis" in *Encyclopedia of Statistical Sciences, Vol 2,* Kotz, S.; Johnson, N. L.; and Read, C. B., Eds., Wiley: New York, NY, 1982, pp. 545–541.
6. Mood, A.; Graybill, F.; and Boes, D. *Introduction to the Theory of Statistics, Third Edition* McGraw-Hill: New York, NY, 1974.
7. Vardeman, S. B. and Jobe, J. M. *Statistical Quality Assurance Methods for Engineers,* Wiley: New York, NY, 1999.

# F

# Simulating False Match Probabilities Based on Normal Theory[1]

## WHY THE FALSE MATCH PROBABILITY DEPENDS ON ONLY RATIO $\delta / \sigma$

As a function of $\delta$, we are interesed in P{match on one element is declared | $|\mu_x - \mu_y| > \delta$}, where $\mu_x$ and $\mu_y$ are the true means of one of the seven elements in the melts of the CS and PS bullets, respectively. The within-replicate variance is generally small, so we assume that the sample means of the three replicates are normally distributed; that is,

$$\bar{x} \sim N(\mu_x, \sigma^2 / 3);\ \bar{y} \sim N(\mu_x + \delta, \sigma^2 / 3),$$

where "~" stands for "is distributed as." Thus, the difference in the means is $\delta$. We further assume that the errors in the measurements leading to $\bar{x}$ and $\bar{y}$ are independent. Based on this specification (or "these assumptions"), statistical theory asserts that

$$(\bar{x} - \bar{y} - \delta) / (\sigma\sqrt{2/3}) \sim N(0,1)$$

$$s_p^2 / \sigma^2 \sim \chi_4^2 / 4$$

where $s_p^2 = (2s_x^2 + 2s_y^2)/4$ and $\chi_4^2$ denotes the chi-squared distribution on four degrees of freedom. If $\sigma^2$ is estimated from a pooled variance on $B$ (more than 2) samples, then $s_p^2/\sigma^2 \sim \chi_{2B}^2/(2B)$. Let $\nu$ equal the number of degrees of freedom used to estimate $\sigma$, for example, $\nu = 4$ if $s_p^2 = (s_x^2 + s_y^2)/2$, or $\nu = 2B$ if $s_p^2 = (s_1^2 +$

[1]Note that the notation used in this appendix differs from that used in the body of the report.

$\ldots + s_B^2)/B$. The ratio of $(\bar{x} - \bar{y} - \delta)$ to $s_p\sqrt{2/3}$ is the same as the distribution of $N(0,1)/\sqrt{\chi_\nu^2/\nu}$, namely a Student's $t_\nu$ ($\nu$ degrees of freedom), so the two-sample $t$ statistic is distributed as a (central) Student's $t$ on $\nu$ degrees of freedom:

$$(\bar{x} - \bar{y} - \delta) / (s_p\sqrt{2/3}) \sim t_4.$$

The FBI criterion for a match on this one element can be written

$$P\{|\bar{x} - \bar{y}| < 2(s_x + s_y) \mid |\mu_x - \mu_y| = \delta\}$$

$$P\{|\bar{x} - \bar{y}|/(s_p\sqrt{2/3}) < 2(s_x + s_y)/(s_p\sqrt{2/3}) \mid |\mu_x - \mu_y| = \delta\}$$

Because $E(s_x) = E(s_y) = 0.8812\sigma$, and $E(s_p) \approx \sigma$ if $\nu > 60$, this reduces very roughly to

$$P\{|\bar{x} - \bar{y}|/(s_p\sqrt{2/3}) < 4.317 \mid |\mu_x - \mu_y| = \delta\}.$$

The approximation is very rough because $E(P\{t < S\}) \neq P\{t < E(S)\}$, where $t$ stands for the two-sample $t$ statistic and $S$ stands for $2(s_x + s_y)/(s_p\sqrt{2/3})$. But it does show that if $\delta$ is very large, this probability is virtually zero (very small false match probability because the probability that the sample means would, by chance, end up very close together is very small). However, if $\delta$ is small, the probability is quite close to 1.

The equivalence $t$ test proceeds as follows. Assume

$$H_0: |\mu_x - \mu_y| \geq \delta$$

$$H_1: |\mu_x - \mu_y| < \delta$$

where $H_0$ is the null hypothesis that the true population means differ by at least $\delta$, and the alternative hypothesis is that they are within $\delta$ of each other. The two-sample $t$ test would reject $H_0$ in favor of $H_1$ if the sample means are too close, that is, if $|\bar{x} - \bar{y}| / (s_p\sqrt{2/3} < K_\alpha(n,\delta)$, where $K_\alpha(n,\delta)$ is chosen so that $P\{|\bar{x} - \bar{y}| /(s_p\sqrt{2/3}) \mid |\mu_x - \mu_y| = \delta\}$ does not exceed a preset per-element risk level of $\alpha$ (in Chapter 3, we used $\alpha = 0.30$). Rewriting that equation, and writing $K_\alpha$ for $K_\alpha(n,\delta)$,

$$P\{-K_\alpha - \delta/(s_p\sqrt{2/3}) \leq (\bar{x} - \bar{y})/(s_p\sqrt{2/3}) \leq K_\alpha - \delta/(s_p\sqrt{2/3})\} = \alpha.$$

When $\nu$ is large $s_p \approx 0$, and therefore the quantity $\delta/(s_p\sqrt{2/3}) \approx 1.2247\delta/\sigma$.

That shows that the false match probability depends on $\delta$ and $\sigma$ only through the ratio. (The argument is a little more complicated when $\nu$ is small, because the ratio $(s_p\sqrt{2/3}/\sigma)$ is a random quantity, but the conclusion will be the same.) Also, when $\nu$ is large, the quantity $(\bar{x}-\bar{y})/(s_p\sqrt{2/3}) \approx (\bar{x}-\bar{y})/(\sigma\sqrt{2/3})$ , which is distributed as a standard normal distribution. So the probability can be written

$$P\{-K_\alpha - \delta/(s_p\sqrt{2/3}) \le Z \le K_\alpha - \delta/(s_p\sqrt{2/3})\} = \alpha$$
$$\Rightarrow \Phi(K_\alpha - \delta/(s_p\sqrt{2/3})) - \Phi(-K_\alpha - \delta/(s_p\sqrt{2/3})) = \alpha$$

where $\Phi(\cdot)$ denotes the standard cumulative normal distribution function (for example, $\Phi(1.645) = 0.95$). So, for large values of $\nu$, the nonlinear equation can be solved for $K_\alpha$, so that the probability of interest does not exceed $\alpha$. For small values of $\nu$, $K_\alpha$ is the $100(1-\alpha)\%$ point of the non-central t distribution with $\nu$ degrees of freedom and noncentrality parameter $\sqrt{n/2}\ \delta/\sigma$ (Ref. 14).

Values of $K_\alpha$ are given in Table F.1 below, for various values of $\alpha$ (0.30, 0.25, 0.20, 0.10, 0.05, 0.01, and 0.0004), degrees of freedom (4, 40, 100, and 200), and $\delta/\sigma$ (0.25, 0.33, 0.50, 1, 1.5, 2, and 3). The theory for Hotelling's $T^2$

TABLE F.1 Values of $K_\alpha(n,\nu)$ Used in Equivalence $t$ Test (Need to Multiply by $\sqrt{2/3}$)

$\alpha = 0.30$, n = 3

| | ($\delta/\sigma$) | | | | | | |
|---|---|---|---|---|---|---|---|
| | 0.25 | 0.33 | 0.50 | 1 | 1.5 | 2 | 3 |
| $\nu = 4$ | 0.43397 | 0.44918 | 0.49809 | 0.81095 | 1.35161 | 1.94726 | 3.12279 |
| 40 | 0.40683 | 0.42113 | 0.46725 | 0.77043 | 1.31802 | 1.92530 | 3.13875 |
| 100 | 0.40495 | 0.41919 | 0.46511 | 0.76783 | 1.31622 | 1.92511 | 3.14500 |
| 200 | 0.40435 | 0.41857 | 0.46443 | 0.76697 | 1.31563 | 1.92510 | 3.14734 |

$\alpha = 0.30$, n = 5

| | ($\delta/\sigma$) | | | | | |
|---|---|---|---|---|---|---|
| | 0.25 | 0.33 | 0.50 | 1 | 2 | 3 |
| $\nu = 4$ | 0.44761 | 0.47385 | 0.56076 | 1.11014 | 2.63496 | 4.12933 |
| 40 | 0.41965 | 0.44436 | 0.52681 | 1.07231 | 2.63226 | 4.19067 |
| 100 | 0.41771 | 0.44232 | 0.52445 | 1.06984 | 2.63546 | 4.20685 |
| 200 | 0.41710 | 0.44167 | 0.52370 | 1.06906 | 2.63664 | 4.21278 |

α = 0.25, n = 3

| | (δ / σ) | | | | | | |
|---|---|---|---|---|---|---|---|
| | 0.25 | 0.33 | 0.50 | 1 | 1.5 | 2 | 3 |
| ν = 4 | 0.35772 | 0.37030 | 0.41092 | 0.68143 | 1.19242 | 1.77413 | 2.91548 |
| 40 | 0.33633 | 0.34818 | 0.38655 | 0.64811 | 1.16900 | 1.77305 | 2.98156 |
| 100 | 0.33484 | 0.34664 | 0.38484 | 0.64578 | 1.16765 | 1.77420 | 2.99223 |
| 200 | 0.33437 | 0.34615 | 0.38430 | 0.64503 | 1.16722 | 1.77461 | 2.99595 |

α = 0.25, n = 5

| | (δ / σ) | | | | | | |
|---|---|---|---|---|---|---|---|
| | 0.25 | 0.33 | 0.50 | 1 | 1.5 | 2 | 3 |
| ν = 4 | 0.36900 | 0.39075 | 0.46350 | 0.95953 | 1.70024 | 2.44328 | 3.88533 |
| 40 | 0.34696 | 0.36748 | 0.43648 | 0.92903 | 1.69596 | 2.47772 | 4.02810 |
| 100 | 0.34542 | 0.36586 | 0.43459 | 0.92698 | 1.69672 | 2.48365 | 4.05178 |
| 200 | 0.34493 | 0.36534 | 0.43399 | 0.92633 | 1.69700 | 2.48570 | 4.06021 |

α = 0.222, n = 3

| | (δ / σ) | | | | | | |
|---|---|---|---|---|---|---|---|
| | 0.25 | 0.33 | 0.50 | 1 | 1.5 | 2 | 3 |
| 4 | 0.31603 | 0.32716 | 0.36318 | 0.60827 | 1.09914 | 1.67316 | 2.79619 |
| 40 | 0.29754 | 0.30804 | 0.34207 | 0.57848 | 1.07949 | 1.68119 | 2.88735 |
| 100 | 0.29625 | 0.30670 | 0.34060 | 0.57638 | 1.07834 | 1.68290 | 2.90000 |
| 200 | 0.29584 | 0.30627 | 0.34013 | 0.57571 | 1.07798 | 1.68350 | 2.90436 |

α = 0.222, n = 5

| | (δ / σ) | | | | | | |
|---|---|---|---|---|---|---|---|
| | 0.25 | 0.33 | 0.50 | 1 | 1.5 | 2 | 3 |
| 3 | 0.32601 | 0.34528 | 0.41003 | 0.87198 | 1.60019 | 2.33249 | 3.74571 |
| 40 | 0.30695 | 0.32514 | 0.38655 | 0.84440 | 1.60422 | 2.38467 | 3.93060 |
| 100 | 0.30562 | 0.32374 | 0.38490 | 0.84252 | 1.60548 | 2.39187 | 3.95822 |
| 200 | 0.30520 | 0.32329 | 0.38438 | 0.84192 | 1.60592 | 2.39434 | 3.96795 |

α = 0.20, n = 3

| | (δ / σ) | | | | | |
|---|---|---|---|---|---|---|
| | 0.25 | 0.33 | 0.50 | 1 | 2 | 3 |
| ν = 4 | 0.28370 | 0.29370 | 0.32612 | 0.55032 | 1.59066 | 2.69968 |
| 40 | 0.26736 | 0.27680 | 0.30744 | 0.52321 | 1.60451 | 2.80887 |
| 100 | 0.26622 | 0.27561 | 0.30613 | 0.52129 | 1.60656 | 2.82294 |
| 200 | 0.26585 | 0.27523 | 0.30571 | 0.52068 | 1.60725 | 2.82774 |

α = 0.20, n = 5

| | (δ / σ) | | | | | |
|---|---|---|---|---|---|---|
| | 0.25 | 0.33 | 0.50 | 1 | 2 | 3 |
| ν = 4 | 0.29266 | 0.30999 | 0.36844 | 0.80094 | 2.24256 | 3.63322 |
| 40 | 0.27582 | 0.29219 | 0.34759 | 0.77521 | 2.30710 | 3.84954 |
| 100 | 0.27464 | 0.29094 | 0.34612 | 0.77341 | 2.31517 | 3.88010 |
| 200 | 0.27426 | 0.29054 | 0.34566 | 0.77285 | 2.31790 | 3.89081 |

α = 0.10, n = 3

*continued*

TABLE F.1 *continued*

| | (δ / σ) | | | | | |
|---|---|---|---|---|---|---|
| | 0.25 | 0.33 | 0.50 | 1 | 2 | 3 |
| ν = 4 | 0.14025 | 0.14521 | 0.16138 | 0.28009 | 1.14311 | 2.19312 |
| 40 | 0.13257 | 0.13726 | 0.15256 | 0.26552 | 1.16523 | 2.36203 |
| 100 | 0.13203 | 0.13670 | 0.15193 | 0.26449 | 1.16738 | 2.38036 |
| 200 | 0.13186 | 0.13653 | 0.15174 | 0.26416 | 1.16808 | 2.38652 |

$\alpha = 0.10$, $n = 5$

| | (δ / σ) | | | | | |
|---|---|---|---|---|---|---|
| | 0.25 | 0.33 | 0.50 | 1 | 2 | 3 |
| ν = 4 | 0.14470 | 0.15332 | 0.18272 | 0.44037 | 1.76516 | 3.05121 |
| 40 | 0.13678 | 0.14493 | 0.17277 | 0.42178 | 1.86406 | 3.39055 |
| 100 | 0.13622 | 0.14434 | 0.17207 | 0.42044 | 1.87408 | 3.43264 |
| 200 | 0.13604 | 0.14416 | 0.17184 | 0.42001 | 1.87741 | 3.44712 |

$\alpha = 0.05$, $n = 3$

| | (δ / σ) | | | | | |
|---|---|---|---|---|---|---|
| | 0.25 | 0.33 | 0.50 | 1 | 2 | 3 |
| 4 | 0.07000 | 0.07241 | 0.08048 | 0.14085 | 0.80000 | 1.82564 |
| 40 | 0.06614 | 0.06847 | 0.07612 | 0.13329 | 0.80877 | 2.00110 |
| 100 | 0.06580 | 0.06812 | 0.07584 | 0.13280 | 0.80951 | 2.01774 |
| 200 | 0.06588 | 0.06822 | 0.07573 | 0.13263 | 0.80976 | 2.02351 |

$\alpha = 0.05$, $n = 5$

| | (δ / σ) | | | | | |
|---|---|---|---|---|---|---|
| | 0.25 | 0.33 | 0.50 | 1 | 2 | 3 |
| 4 | 0.07215 | 0.07645 | 0.09118 | 0.22900 | 1.41106 | 2.64066 |
| 40 | 0.06825 | 0.07232 | 0.08626 | 0.21748 | 1.50372 | 3.02532 |
| 100 | 0.06798 | 0.07203 | 0.08591 | 0.21672 | 1.51184 | 3.06786 |
| 200 | 0.06789 | 0.07194 | 0.08580 | 0.21647 | 1.51462 | 3.08296 |

$\alpha = 0.01$, $n = 3$

| | (δ / σ) | | | | | |
|---|---|---|---|---|---|---|
| | 0.25 | 0.33 | 0.50 | 1 | 2 | 3 |
| 4 | 0.01397 | 0.01447 | 0.01608 | 0.02823 | 0.25124 | 1.21164 |
| 40 | 0.01322 | 0.01369 | 0.01522 | 0.02671 | 0.24129 | 1.33049 |
| 100 | 0.01317 | 0.01364 | 0.01516 | 0.02660 | 0.24062 | 1.34080 |
| 200 | 0.01315 | 0.01352 | 0.01514 | 0.02656 | 0.24040 | 1.34432 |

$\alpha = 0.01$, $n = 5$

| | (δ / σ) | | | | | |
|---|---|---|---|---|---|---|
| | 0.25 | 0.33 | 0.50 | 1 | 2 | 3 |
| 4 | 0.01442 | 0.01528 | 0.01823 | 0.04651 | 0.79664 | 1.98837 |
| 40 | 0.01364 | 0.01446 | 0.01724 | 0.04400 | 0.83240 | 2.35173 |
| 100 | 0.01359 | 0.01440 | 0.01717 | 0.04383 | 0.83521 | 2.38989 |
| 200 | 0.01357 | 0.01438 | 0.01715 | 0.04378 | 0.83616 | 2.40330 |

$\alpha = 0.0004$, n = 3

| | (δ / σ) | | | | | | |
|---|---|---|---|---|---|---|---|
| | 0.25 | 0.33 | 0.50 | 1 | 2 | 3 | 4.4 |
| 4 | 0.00056 | 0.00058 | 0.00064 | 0.00113 | 0.01071 | 0.34213 | 1.5877 |
| 40 | 0.00053 | 0.00055 | 0.00061 | 0.00107 | 0.01013 | 0.34139 | 1.9668 |
| 100 | 0.00053 | 0.00055 | 0.00061 | 0.00107 | 0.01009 | 0.34133 | 2.0072 |
| 200 | 0.00053 | 0.00055 | 0.00060 | 0.00106 | 0.01008 | 0.34131 | 2.0215 |

$\alpha = 0.0004$, n = 5

| | (δ / σ) | | | | | |
|---|---|---|---|---|---|---|
| | 0.25 | 0.33 | 0.50 | 1 | 2 | 3 |
| 4 | 0.00057 | 0.00061 | 0.00073 | 0.00186 | 0.07825 | 1.16693 |
| 40 | 0.00055 | 0.00058 | 0.00069 | 0.00176 | 0.07424 | 1.36013 |
| 100 | 0.00054 | 0.00057 | 0.00069 | 0.00175 | 0.07397 | 1.37811 |
| 200 | 0.00054 | 0.00057 | 0.00068 | 0.00175 | 0.07389 | 1.38431 |

Note: In each subtable, the row corresponds to different values of $\nu$ = number of degrees of freedom used in $s_p$ to estimate $\sigma$ (number of bullets = $\nu / 2 + 1$ with two measurements per bullet).

is similar (it uses vectors and matrices instead of scalars), and the resulting critical value comes from a noncentral F distribution (Ref. 15).[2]

## ESTIMATING MEASUREMENT UNCERTAINTY WITH POOLED STANDARD DEVIATIONS

Chapter 3 states that a pooled estimate of the measurement uncertainty $\sigma$, $s_p$, is more accurate and precise than an estimate based on only $s_x$, the sample SD based on only three normally distributed measurements. That statement follows from the fact that a squared sample SD has a chi-squared distribution; specifically, $(n-1)s^2 / \sigma^2$ has a chi-squared distribution on $(n-1)$ degrees of freedom, where $s$ is based on $n$ observations. The mean of the square root of a chi-squared random variable based on $\nu = (n-1)$ degrees of freedom is $\sqrt{2}\Gamma((\nu+1)/2)/\Gamma(\nu/2)$, where $\Gamma(\cdot)$ is the gamma function. For $\nu = (n-1) = 2$, $E(s) = 0.8812\sigma$; for $\nu = 4$ (i.e., estimating $\sigma$ by $\sqrt{(s_x^2 + s_y^2)/2)}$), $E(s) = 0.9400\sigma$; for $\nu = 200$ (that is, estimating $\sigma$ by the square root of the mean of the squared SDs from 100 bullets), $E(s) \approx \sigma$. In addition, the probability that $s$ exceeds $1.25\sigma$ when $n = 2$ (that is, using only one bullet) is 0.21 but falls to 0.00028 when $\nu = 200$. For those

[2]These values were determined by using a simple binary search algorithm for the value $\alpha$ and the R function pf($x$, 1, dof, 0.5*$n$*$E$), where $n = 3$ or 5 and E = $(\delta / \sigma)^2$. R is a statistical-analysis software program that is downloadable from http://www.r-project.org.

reasons, $s_p$ based on many bullets is preferable to estimating $\sigma$ by using only three measurements on a single bullet.

## WITHIN-BULLET VARIANCES, COVARIANCES, AND CORRELATIONS FOR FEDERAL BULLET DATA SET

The data on the Federal bullets contained measurements on six of the seven elements (all but Cd) with ICP-OES. They allowed estimation of within-bullet variances, covariances, and correlations among the six elements. According to the formula in Appendix K, now applied to the six elements, the estimated within-bullet variance matrix is given below. The correlation matrix is found in the usual way (for example, Cor (Ag, Sb) = Covariance(Ag,Sb)/[SD(Ag) SD(Sb)]. Covariances and correlations between Cd and all other elements are assumed to be zero. The correlation matrix was used to demonstrate the use of the equivalence Hotelling's $T^2$ test. Because it is based on 200 bullets measured in 1991, it is presented here for illustrative purposes only.

Within-Bullet Variances and Covariances $\times 10^5$, log(Federal Data)

| | ICP-As | ICP-Sb | ICP-Sn | ICP-Bi | ICP-Cu | ICP-Ag |
|---|---|---|---|---|---|---|
| ICP-As | 187 | 27 | 31 | 31 | 37 | 77 |
| ICP-Sb | 20 | 37 | 25 | 18 | 25 | 39 |
| ICP-Sn | 31 | 25 | 106 | 16 | 29 | 41 |
| ICP-Bi | 31 | 18 | 16 | 90 | 14 | 44 |
| ICP-Cu | 37 | 25 | 29 | 14 | 40 | 42 |
| ICP-Ag | 77 | 39 | 41 | 44 | 42 | 681 |

Within-Bullet Correlations, Federal Data

| | ICP-As | ICP-Sb | ICP-Sn | ICP-Bi | ICP-Cu | ICP-Ag | (Cd) |
|---|---|---|---|---|---|---|---|
| ICP-As | 1.000 | 0.320 | 0.222 | 0.236 | 0.420 | 0.215 | 0.000 |
| ICP-Sb | 0.320 | 1.000 | 0.390 | 0.304 | 0.635 | 0.242 | 0.000 |
| ICP-Sn | 0.222 | 0.390 | 1.000 | 0.163 | 0.440 | 0.154 | 0.000 |
| ICP-Bi | 0.236 | 0.304 | 0.163 | 1.000 | 0.240 | 0.179 | 0.000 |
| ICP-Cu | 0.420 | 0.635 | 0.440 | 0.240 | 1.000 | 0.251 | 0.000 |
| ICP-Ag | 0.215 | 0.242 | 0.154 | 0.179 | 0.251 | 1.000 | 0.000 |
| (Cd) | 0.000 | 0.000 | 0.000 | 0.000 | 0.000 | 0.000 | 1.000 |

## BETWEEN-ELEMENT CORRELATIONS

In Chapter 3, correlations between mean concentrations of bullets were estimated by using the Pearson correlation coefficient (see equation 2). One reviewer suggested that Spearman's rank correlation may be more appropriate, as it provides a nonparametric estimate of the monotonic association between two variables. Spearman's rank correlation coefficient takes the same form as Equation 2, but with the *ranks* of the values (numbers 1, 2, 3, ... , $n$ = number of data pairs) rather than values themselves. The table below consists of 49 entries, corresponding to all possible pairs of the seven elements. The value 1.000 on the diagonal confirms a correlation of 1.000 for an element with itself. The values in the cells on either side of the diagonal are the same because the correlation between, say, As and Sb is the same as that between Sb and As. For these off-diagonal cells, the first line reflects the conventional Pearson correlation coefficient based on the 1,373-bullet subset from the 1,837-bullet subset (bullets with all seven measured elements or with six measured and one imputed for Cd). The second line is Spearman's rank correlation coefficient on rank(data), again for

| | | |
|---|---|---|
| Line 1: | conventional correlations on log(data), | 1,373-bullet subset |
| Line 2: | Spearman correlations on rank(data), | 1,373-bullet subset |
| Line 3: | Spearman correlations on rank(data), | 1,837-bullet subset |
| Line 4: | Number of pairs in Spearman correlation, | 1,837-bullet subset |

(Note: 1.000 on the diagonal is indicated on line 1 only)

| | As | Sb | Sn | Bi | Cu | Ag | Cd |
|---|---|---|---|---|---|---|---|
| As | 1.000 | 0.556 | 0.624 | 0.148 | 0.388 | 0.186 | 0.242 |
| | | 0.697 | 0.666 | 0.165 | 0.386 | 0.211 | 0.166 |
| | | 0.678 | 0.667 | 0.178 | 0.392 | 0.216 | 0.279 |
| | | 1750 | 1,381 | 1742 | 1,743 | 1,750 | 856 |
| Sb | 0.556 | 1.000 | 0.455 | 0.157 | 0.358 | 0.180 | 0.132 |
| | 0.697 | | 0.556 | 0.058 | 0.241 | 0.194 | 0.081 |
| | 0.678 | | 0.560 | 0.054 | 0.233 | 0.190 | 0.173 |
| | 1,750 | | 1,387 | 1829 | 1,826 | 1,837 | 857 |
| Sn | 0.624 | 0.455 | 1.000 | 0.176 | 0.200 | 0.258 | 0.178 |
| | 0.666 | 0.556 | | 0.153 | 0.207 | 0.168 | 0.218 |
| | 0.667 | 0.560 | | 0.152 | 0.208 | 0.165 | 0.385 |
| | 1,381 | 1,387 | | 1385 | 1380 | 1387 | 857 |
| Bi | 0.148 | 0.157 | 0.176 | 1.000 | 0.116 | 0.560 | 0.030 |
| | 0.165 | 0.058 | 0.153 | | 0.081 | 0.499 | 0.103 |
| | 0.178 | 0.054 | 0.152 | | 0.099 | 0.522 | 0.165 |
| | 1,742 | 1,829 | 1,385 | | 1,818 | 1,829 | 857 |
| Cu | 0.388 | 0.358 | 0.200 | 0.116 | 1.000 | 0.258 | 0.111 |
| | 0.386 | 0.241 | 0.207 | 0.081 | | 0.206 | 0.151 |
| | 0.392 | 0.233 | 0.208 | 0.099 | | 0.260 | 0.115 |
| | 1,743 | 1,826 | 1,380 | 1818 | | 1826 | 855 |
| Ag | 0.186 | 0.180 | 0.258 | 0.560 | 0.258 | 1.000 | 0.077 |
| | 0.211 | 0.194 | 0.168 | 0.499 | 0.206 | | 0.063 |
| | 0.216 | 0.190 | 0.165 | 0.522 | 0.260 | | 0.115 |
| | 1,750 | 1,837 | 1,387 | 1829 | 1,826 | | 857 |
| Cd | 0.242 | 0.132 | 0.178 | 0.030 | 0.111 | 0.077 | 1.000 |
| | 0.166 | 0.081 | 0.218 | 0.103 | 0.151 | 0.063 | |
| | 0.279 | 0.173 | 0.385 | 0.165 | 0.251 | 0.115 | |
| | 857 | 857 | 857 | 857 | 855 | 857 | |

the 1,373-bullet subset. The third line is Spearman's rank correlation coefficient on the entire 1,837-bullet subset (some bullets had only three, four, five, or six elements measured). The fourth line gives the number of pairs in Spearman's rank correlation coefficient calculation. All three sets of correlation coefficients are highly consistent with each other. Regardless of the method used to estimate the linear association between elements, associations between As and Sb, between As and Sn, between Sb and Sn, and between Ag and Bi are rather high. Because the 1,837-bullet subset is not a random sample from any population, we refrain from stating a level of "significance" for these values, noting only that regardless of the method used to estimate the linear association between elements, associations between As and Sb, between As and Sn, between Sb and Sn, and between Ag and Bi are higher than those for the other 17 pairs of elements.

# G

# Data Analysis of Table 1, Randich et al.

The Randich et al. (Ref. 1) paper is based on an analysis of compositional data provided by two secondary lead smelters to bullet manufacturers on their lead alloy shipments. For each element, Randich et al. provide three measurements from each of 28 lead (melt) lots being poured into molds. The measurements were taken at the beginning (B), middle (M), and end (E) "position" of each pour. In this appendix, the variability in the measurements within a lot (due to position) is compared with the variability across lots. Consistent patterns in the lots and positions are also investigated.

Let $u_{ijk}$ denote the logarithm of the reported value in position $i$ ($i = 1, 2, 3$, for B, M, E) in lot $j$ ($j = 1, \ldots, 28$), on element $k$ ($k = 1, \ldots, 6$, for Sb, Sn, Cu, As, Bi, and Ag). A simple additive model for $u_{ijk}$ in terms of the two factors position and lot is

$$u_{ijk} = \phi_k + \rho_{ik} + \lambda_{jk} + \varepsilon_{ijk},$$

where $\phi_k$ denotes the typical value of $u_{ijk}$ over all positions and lots (usually estimated as the mean over all positions and lots, $\hat{\phi}_k = \bar{u}_{\cdot\cdot k}$); $\rho_{ik}$ denotes the typical effect of position $i$ for element $k$, above or below $\phi_k$ (usually estimated as the mean over all lots minus the overall mean, $\hat{\rho}_{ik} = \bar{u}_{i\cdot k} - \bar{u}_{\cdot\cdot k}$); $\lambda_{jk}$ denotes the typical effect of lot $j$ for element $k$, above or below $\phi_k$ (usually estimated as the mean over all positions minus the overall mean, $\hat{\lambda}_{jk} = \bar{\mu}_{\cdot jk} - \bar{u}_{\cdot\cdot k}$); and $\varepsilon_{ijk}$ is the error term that accounts for any difference that remains between $u_{ijk}$ and the sum of the effects just defined (usually estimated as

$$\hat{\varepsilon}_{ijk} = u_{ijk} - [\bar{u}_{\cdot\cdot k} + (\bar{u}_{i\cdot k} - \bar{u}_{\cdot\cdot k}) + (\bar{u}_{\cdot jk} - \bar{u}_{\cdot\cdot k})] = u_{ijk} - \bar{u}_{i\cdot k} - \bar{u}_{\cdot jk} + \bar{u}_{\cdot\cdot k}).$$

Because replicate measurements are not included in Table 1 of Randich et al., we are unable to assess the existence of an interaction term between position and lot; such an interaction, if it exists, must be incorporated into the error term, which also includes simple measurement error. The parameters of the model ($\phi_k$, $\rho_{ik}$, $\lambda_{jk}$) can also be estimated more robustly via median polish (Ref. 2), which uses medians rather than means and thus provides more robust estimates, particularly when the data include a few outliers or extreme values that will adversely affect sample means (but not sample medians). This additive model was verified for each element by using Tukey's diagnostic plot for two-way tables (Ref. 2, 3).

The conventional way to assess the signficance of the two factors is to compare the variance of the position effects, $\text{Var}(\hat{\rho}_{ik})$, and the variance of the lot effects, $\text{Var}(\hat{\lambda}_{jk})$, scaled to the level of a single observation, with the variance of the estimated error term, $\text{Var}(r_{ijk})$. Under the null hypothesis that all $\rho_{ik}$ are zero (position has no particular effect on the measurements, beyond the anticipated measurement error), the ratio of $28 \cdot \text{Var}(\hat{\rho}_{ik})$ to $\text{Var}(\hat{\varepsilon}_{ijk})$ should follow an F distribution with two and 54 degrees of freedom; ratios that exceed 3.168 would be evidence that position affects measurements more than could be expected from mere measurement error.

Table G.1 below provides the results of the two-way analysis of variance with two factors, position and lot, for each element. The variances of the effects, scaled to the level of a single observation, are given in the column headed "Mean Sq"; the ratio of the mean squares is given under "F Value"; and the P value of

TABLE G.1 Analyses of Variance for Log(Measurement) Using Table 1 in Randich et al. (Ref. 1)

| Sb | Df | Sum Sq | Mean Sq | F Value | Pr (> F) | MS (median polish) |
|---|---|---|---|---|---|---|
| Position | 2 | 0.001806 | 0.000903 | 2.9449 | 0.06111 | 0.004 |
| Lot | 27 | 0.111378 | 0.004125 | 13.4514 | 1.386e-15 | 0.0042 |
| Residuals | 54 | 0.016560 | 0.000307 | | | |

| Sn | Df | Sum Sq | Mean Sq | F Value | Pr (> F) | MS (median polish) |
|---|---|---|---|---|---|---|
| Position | 2 | 2.701 | 1.351 | 7.5676 | 0.001267 | 0.2345 |
| Lot | 27 | 147.703 | 5.470 | 30.6527 | <2.2e-16 | 6.0735 |
| Residuals | 54 | 9.637 | 0.178 | | | |

| Cu | Df | Sum Sq | Mean Sq | F Value | Pr (> F) | MS (median polish) |
|---|---|---|---|---|---|---|
| Position | 2 | 0.006 | 0.003 | 0.1462 | 0.8643 | 0.00003 |
| Lot | 27 | 102.395 | 3.792 | 176.9645 | <2e-16 | 4.1465 |
| Residuals | 54 | 1.157 | 0.021 | | | |

TABLE G.1 *continued*

| As | Df | Sum Sq | Mean Sq | F Value | Pr (> F) | MS (median polish) |
|---|---|---|---|---|---|---|
| Position | 2 | 0.0127 | 0.0063 | 2.1046 | 0.1318 | 0.0036 |
| Lot | 27 | 15.4211 | 0.5712 | 189.5335 | <2e-16 | .5579 |
| Residuals | 54 | 0.1627 | 0.0030 | | | |

| Bi | Df | Sum Sq | Mean Sq | F Value | Pr (> F) | MS (median polish) |
|---|---|---|---|---|---|---|
| Position | 2 | 0.000049 | 0.000024 | 0.3299 | 0.7204 | 0.0000 |
| Lot | 27 | 0.163701 | 0.006063 | 81.9890 | <2e-16 | 0.0061 |
| Residuals | 54 | 0.003993 | 0.000074 | | | |

| Ag | Df | Sum Sq | Mean Sq | F Value | Pr (> F) | MS (median polish) |
|---|---|---|---|---|---|---|
| Position | 2 | 0.00095 | 0.00047 | 1.6065 | 0.21 | 0.0000 |
| Lot | 27 | 1.95592 | 0.07244 | 245.6707 | <2e-16 | 0.0735 |
| Residuals | 54 | 0.01592 | 0.00029 | | | |

this statistic is listed under "Pr(> F)". For comparison, the equivalent mean square under the median polish analysis is also given; notice that, for the most part, the values are consistent with the mean squares given by the conventional analysis of variance, except for Sn, for which the mean square for position is almost 6 times smaller under the median polish (1.351 versus 0.2345).

Only for Sn did the ratio of the mean square for position (B, M, E) to the residual mean square exceed 3.168 (1.351/0.178); for all other elements, this ratio was well below this critical point. (The significance for Sn may have come from the nonrobustness of the sample means caused by two unusually low values: Lot #424, E = 21 (B = 414, M = 414); and Lot #454, E = 45 (B = 377, M = 367). When using median polish as the analysis rather than conventional analysis of variance, the ratio is (0.2345/0.178) = 1.317 (not significant).) For all elements, the effect of lot is highly significant; differences among lots characterize nearly all the variability in these data for all elements.

Table G.2 provides the estimates of the position and lot effects in this format:

<table>
<tr><td colspan="2" rowspan="2"></td><td colspan="5">Lot Number</td><td></td></tr>
<tr><td>1</td><td>2</td><td>3</td><td>...</td><td>28</td><td>Row Effect</td></tr>
<tr><td rowspan="3">Position</td><td>B</td><td colspan="5" rowspan="3">Residual</td><td rowspan="3"></td></tr>
<tr><td>M</td></tr>
<tr><td>E</td></tr>
<tr><td colspan="2">Lot Effect</td><td></td><td></td><td></td><td></td><td></td><td>Overall Effect</td></tr>
</table>

The analysis suggests that the variation observed in the measurements at different positions is not significantly larger than that observed from the analytical measurement error. All analyses were conducted with the statistics package R (Ref. 4).

TABLE G.2 Median Polish on Logarithms (Results Multiplied by 1,000 to Avoid Decimal Points)

| **Sb** | **423** | **424** | **425** | **426** | **427** | **429** | **444** | **445** | **446** | **447** | **448** |
|---|---|---|---|---|---|---|---|---|---|---|---|
| **1** | −7 | 0 | −4 | −10 | 6 | 0 | 19 | 7 | 1 | −15 | 0 |
| **2** | 0 | 0 | 0 | 0 | −3 | −1 | 0 | −3 | 0 | 1 | 3 |
| **3** | 9 | −104 | 2 | 24 | 0 | 6 | −5 | 0 | −8 | 0 | −5 |
| **Column Effect** | −40 | 6 | 12 | 27 | −56 | 57 | 34 | −53 | 1 | 13 | 38 |

| | **450** | **451** | **452** | **453** | **454** | **455** | **456** | **457** | **458** | **459** | **460** |
|---|---|---|---|---|---|---|---|---|---|---|---|
| **1** | −10 | −1 | −3 | 0 | 0 | 0 | 0 | −2 | 0 | −5 | −4 |
| **2** | 0 | 0 | 0 | 1 | 8 | −4 | −9 | 2 | 3 | 0 | 0 |
| **3** | 3 | 11 | 8 | −48 | −33 | 12 | 5 | 0 | −3 | 2 | 44 |
| **Column Effect** | −16 | −35 | −9 | −1 | 57 | −53 | −34 | 47 | −49 | 52 | −12 |

| | **461** | **463** | **464** | **465** | **466** | **467** | **Row Effect** |
|---|---|---|---|---|---|---|---|
| **1** | 66 | 0 | 0 | 1 | 0 | 4 | 0 |
| **2** | −5 | −5 | −4 | 0 | −8 | 0 | 0 |
| **3** | 0 | 5 | 0 | −21 | 10 | −2 | −6 |
| **Column Effect** | −32 | 53 | −34 | −37 | 23 | 1 | 6559 |

| **Sn** | **423** | **424** | **425** | **426** | **427** | **429** | **444** | **445** | **446** | **447** | **448** |
|---|---|---|---|---|---|---|---|---|---|---|---|
| **1** | 0 | 0 | 0 | −41 | 144 | −45 | 271 | 0 | 0 | 0 | −179 |
| **2** | 127 | 69 | −27 | 0 | −192 | 0 | 0 | 4 | 61 | −55 | 0 |
| **3** | −120 | −2800 | 11 | 148 | 0 | 60 | −53 | −42 | −15 | 168 | 9 |
| **Column Effect** | −1050 | 371 | −625 | 672 | −2909 | 1442 | −659 | −408 | −884 | −618 | 108 |

| | **450** | **451** | **452** | **453** | **454** | **455** | **456** | **457** | **458** | **459** | **460** |
|---|---|---|---|---|---|---|---|---|---|---|---|
| **1** | 0 | 605 | −22 | 1428 | 0 | −45 | −6 | 240 | 41 | −77 | −5 |
| **2** | −9 | 0 | 0 | −112 | 42 | 0 | 28 | −30 | 0 | 0 | 0 |
| **3** | 201 | −313 | 83 | 0 | −1944 | 99 | 0 | 0 | −176 | 88 | 139 |
| **Column Effect** | −122 | −2328 | −942 | −5474 | 277 | 338 | 203 | −1067 | −349 | 849 | 787 |

| | **461** | **463** | **464** | **465** | **466** | **467** | **Row Effect** |
|---|---|---|---|---|---|---|---|
| **1** | −22 | −65 | 0 | 436 | 0 | −54 | 69 |
| **2** | 0 | 0 | 53 | −71 | −4 | 0 | 0 |
| **3** | 118 | 112 | −443 | 0 | 95 | 68 | −112 |
| **Column Effect** | 908 | 933 | 938 | −117 | 846 | 560 | 5586 |

Two unusual residuals:
Lot #424, “E” = 21 (B = 414, M = 414)
Lot #454, “E” = 45 (B = 377, M = 367)

TABLE G.2 *continued*

| **Cu** | **423** | **424** | **425** | **426** | **427** | **429** | **444** | **445** | **446** | **447** | **448** |
|---|---|---|---|---|---|---|---|---|---|---|---|
| **1** | –166 | –19 | –18 | 93 | –2 | –13 | 0 | –8 | 0 | 0 | 106 |
| **2** | 0 | 0 | 0 | 0 | 0 | 0 | 2 | 0 | 35 | 34 | –23 |
| **3** | 12 | 51 | 0 | –121 | 0 | 0 | –38 | 0 | –43 | –21 | 0 |
| **Column Effect** | 607 | 258 | –94 | 418 | 80 | –424 | 436 | 269 | 441 | 307 | –1106 |

| | **450** | **451** | **452** | **453** | **454** | **455** | **456** | **457** | **458** | **459** | **460** |
|---|---|---|---|---|---|---|---|---|---|---|---|
| **1** | –16 | –27 | –37 | 44 | 0 | 27 | 76 | 13 | 0 | –53 | –2 |
| **2** | 0 | 0 | 0 | 0 | 52 | –5 | 0 | 0 | 2 | 0 | 0 |
| **3** | 0 | 24 | 0 | 0 | –470 | 0 | 0 | 0 | –5 | 49 | 288 |
| **Column Effect** | 30 | –495 | –1523 | –30 | 630 | 448 | 330 | 30 | 50 | –1894 | –2405 |

| | **461** | **463** | **464** | **465** | **466** | **467** | **Row Effect** |
|---|---|---|---|---|---|---|---|
| **1** | –2 | 691 | 0 | –242 | 13 | –24 | 2 |
| **2** | 0 | 0 | –28 | 10 | –31 | 0 | 0 |
| **3** | 19 | 0 | 857 | 0 | 0 | 11 | 0 |
| **Column Effect** | –958 | –4890 | –1365 | –255 | –700 | –357 | |

| **As** | **423** | **424** | **425** | **426** | **427** | **429** | **444** | **445** | **446** | **447** | **448** |
|---|---|---|---|---|---|---|---|---|---|---|---|
| **1** | –166 | –19 | –18 | 93 | –2 | –13 | 0 | –8 | 0 | 0 | 106 |
| **2** | 0 | 0 | 0 | 0 | 0 | 0 | 2 | 0 | 35 | 34 | –23 |
| **3** | 12 | 51 | 0 | –121 | 0 | 0 | –38 | 0 | –43 | –21 | 0 |
| **Column Effect** | 607 | 258 | –94 | 418 | 80 | –424 | 436 | 269 | 441 | 307 | –1106 |

| | **450** | **451** | **452** | **453** | **454** | **455** | **456** | **457** | **458** | **459** | **460** |
|---|---|---|---|---|---|---|---|---|---|---|---|
| **1** | –16 | –27 | –37 | 44 | 0 | 27 | 76 | 13 | 0 | –53 | –2 |
| **2** | 0 | 0 | 0 | 0 | 52 | –5 | 0 | 0 | 2 | 0 | 0 |
| **3** | 0 | 24 | 0 | 0 | –470 | 0 | 0 | 0 | –5 | 49 | 288 |
| **Column Effect** | 30 | –495 | –1523 | –30 | 630 | 448 | 330 | 30 | 50 | –1894 | –2405 |

| | **461** | **463** | **464** | **465** | **466** | **467** | **Row Effect** |
|---|---|---|---|---|---|---|---|
| **1** | –2 | 691 | 0 | –242 | 13 | –24 | 2 |
| **2** | 0 | 0 | –28 | 10 | –31 | 0 | 0 |
| **3** | 19 | 0 | 857 | 0 | 0 | 11 | 0 |
| **Column Effect** | –958 | –4890 | –1365 | –255 | –700 | –357 | 4890 |

| **Bi** | **423** | **424** | **425** | **426** | **427** | **429** | **444** | **445** | **446** | **447** | **448** |
|---|---|---|---|---|---|---|---|---|---|---|---|
| **1** | 0 | –11 | 0 | 0 | 10 | –10 | 0 | 10 | 0 | 0 | 0 |
| **2** | –10 | 0 | 0 | 0 | 0 | 0 | 0 | 0 | 0 | 9 | 0 |
| **3** | 0 | 0 | 0 | 10 | 0 | 0 | 0 | 0 | 0 | 0 | 0 |
| **Column Effect** | –5 | –78 | –46 | –25 | –25 | –35 | 15 | 15 | 63 | 90 | 15 |

| | **450** | **451** | **452** | **453** | **454** | **455** | **456** | **457** | **458** | **459** | 460 |
|---|---|---|---|---|---|---|---|---|---|---|---|
| **1** | 0 | –9 | 0 | 52 | 0 | 0 | 0 | 0 | 0 | 0 | 0 |
| **2** | –9 | 0 | 10 | 0 | 0 | –11 | 0 | 0 | 0 | 0 | 0 |
| **3** | 0 | 9 | 0 | –11 | –21 | 0 | 11 | 0 | 0 | 10 | 10 |
| **Column Effect** | 53 | 90 | –25 | –67 | –35 | –67 | –67 | 34 | 25 | 34 | 15 |

TABLE G.2 *continued*

| | **461** | **463** | **464** | **465** | **466** | **467** | **Row Effect** |
|---|---|---|---|---|---|---|---|
| **1** | −10 | 0 | 0 | 0 | 0 | 0 | 0 |
| **2** | 10 | 0 | −10 | 0 | 0 | 0 | 0 |
| **3** | 0 | 0 | 10 | 0 | 10 | 0 | 0 |
| **Column Effect** | −35 | −15 | 5 | 15 | −5 | 5 | 4160 |

| **Ag** | **423** | **424** | **425** | **426** | **427** | **429** | **444** | **445** | **446** | **447** | **448** |
|---|---|---|---|---|---|---|---|---|---|---|---|
| **1** | −166 | −19 | −18 | 93 | −2 | −13 | 0 | −8 | 0 | 0 | 106 |
| **2** | 0 | 0 | 0 | 0 | 0 | 0 | 2 | 0 | 35 | 34 | −23 |
| **3** | 12 | 51 | 0 | −121 | 0 | 0 | −38 | 0 | −43 | −21 | 0 |
| **Column Effect** | 607 | 258 | −94 | 418 | 80 | −424 | 436 | 269 | 441 | 307 | −1106 |

| | **450** | **451** | **452** | **453** | **454** | **455** | **456** | **457** | **458** | **459** | **460** |
|---|---|---|---|---|---|---|---|---|---|---|---|
| **1** | −16 | −27 | −37 | 44 | 0 | 27 | 76 | 13 | 0 | −53 | −2 |
| **2** | 0 | 0 | 0 | 0 | 52 | −5 | 0 | 0 | 2 | 0 | 0 |
| **3** | 0 | 24 | 0 | 0 | −470 | 0 | 0 | 0 | −5 | 49 | 19 |
| **Column Effect** | 30 | −495 | −1523 | −30 | 630 | 448 | 330 | 30 | 50 | −1894 | −958 |

| | **461** | **463** | **464** | **465** | **466** | **467** | **Row Effect** |
|---|---|---|---|---|---|---|---|
| **1** | −2 | 691 | 0 | −242 | 13 | −24 | 2 |
| **2** | 0 | 0 | −28 | 10 | −31 | 0 | 0 |
| **3** | 19 | 0 | 857 | 0 | 0 | 11 | 0 |
| **Column Effect** | −958 | −4890 | −1365 | −255 | −700 | −357 | 4890 |

Note: Lot numbers are given in bold across the top row and 1, 2, and 3 refer to sample's position in lot (beginning, middle, or end).

## REFERENCES

1. Randich, E.; Duerfeldt, W.; McLendon, W.; and Tobin, W. *Foren. Sci. Int.* **2002**,127, 174–191.
2. Tukey, J. W. *Exploratory Data Analysis;* Addison-Wesley: Reading, MA, 1977.
3. Mosteller, F. and Tukey, J. W. *Data Analysis and Regression: A Second Course in Statistics*; Addison-Wesley: Reading, MA, 1977, pp 192–199.
4. R. Copyright 2002, The R Development Core Team, Version 1.5.1 (2002-06-17), for the Linux operating system see *<http://www.r-project.org>*.

# H

# Principal Components Analysis: How Many Elements Should Be Measured?

The number of elements in bullet lead that have been measured has ranged from three to seven, and sometimes the concentration of a measured element is so small as to be undetectable. The optimal number of elements to measure is unclear. An unambiguous way to determine it is to calculate, using two-sample equivalence *t* tests, the probability of a false match on the 1,837-bullet data set as described in Chapter 3. Recall that the equivalence *t* test requires specification of a value $\delta/RE$ where $RE$ = relative error and a value $\alpha$ denoting the expected probability of a false match. Each simulation run would use a different combination of the elements: there are 35 possible subsets of three of the seven elements, 35 possible subsets of four of the seven elements, 21 possible subsets of five of the seven elements, seven possible subsets of six of the seven elements, and one simulation run corresponding to using all seven elements. Among the three-element subsets, the subset with the lowest false match probability would be selected, and a similar process would occur for the four-, five-, and six-element subsets. One could then plot the false match probability as a function of $\delta/RE$ for various choices of $\delta/RE$ and determine the reduction in false match probability in moving from three to seven elements for testing purposes. Such a calculation may well differ if applied to the full (71,000-bullet) data set.

An alternative, easier to apply but less direct approach is to characterize the variability among the bullets using all seven elements. To avoid the problem of many missing values of elemental concentrations in the 1,837-bullet dataset, we will use the 1,373-bullet subset, for which all 7 elemental calculations exist (after inputing some values for Cd). The variability can then be compared with the variability obtained using all possible three-, four-, five-, and six-element subsets. It is likely that the false match probability will be higher in subsets that comprise lesser amounts of the total variability and lower in subsets that comprise nearly all of the variability in the data set. Variability can be characterized by using principal components analysis (PCA).

Consider, for example, a PCA using the first three elements (As, Sb, and Sn—elements "123"), which yields 104.564 as the total variation in the data. PCA provides the three linear combinations that decompose this variation of 104.564 into three linear combinations of the three elements in a sequential fashion: the first linear combination explains the most variation (76.892); the second, independent of the first, explains the next-most (19.512), and the third accounts for the remainder (8.16). The total variation in all seven elements is 136.944. Thus, this three-element subset accounts for (104.564/136.944) × 100%, or 76.3% of the total variation. The results of PCA on all 35 3-element subsets are shown Table H.1; they illustrate that subset "237" (Sb, Sn, and Cd) appears to be best for characterizing the total variability in the set, accounting for (114.503/136.944) × 100% = 83.6% of the variability. Subset "137" (As, Sn, and Cd) is almost as good at (113.274/136.944) × 100% = 83.0%.

PCA is then applied to all 35 possible four-element subsets; the one that accounts for the most variation, (131.562/136.944) × 100% = 96.1%, is subset "1237" (As, Sb, Sn, and Cd). Among the five-element subsets, subset "12357" (As, Sn, Sb, Cu, and Cd) explains the greatest proportion of the variance: (134.419/136.944) × 100% = 98.2%, or about 2.1% more than the subset without Cu. The five-element subset containing Bi instead of Cu is nearly as efficient: (133.554/136.944) × 100% = 97.5%. Finally, among the six-element subsets, "123457" (all but Ag) comes very close to explaining the variation using all seven elements: (136.411/136.944) × 100% = 99.6%. Measuring all elements except Bi is nearly as efficient, explaining (134.951/136.944) × 100% = 98.5% of total variation. The values obtained for each three-, four-, five-, six-, and seven-element subset PCA are found in Tables H.1, H.3, H.5, H.7, and H.9 below. The corresponding variances in order of increasing percentages are found in Tables H.2, H.4, H.6, and H.8.

This calculation may not directly relate to results obtained by simulating the false match probability as described above, but it does give some indication of the contribution of the different elements, and the results appear to be consistent with the impressions of the scientists who have been measuring bullets and making comparisons (Ref. 1-3).

TABLE H.1 Principal Components Analysis on All Three-Element Subsets of 1,373-Bullet Subset. Elements 1, 2, 3, 4, 5, 6, and 7 are As, Sb, Sn, Cu, Bi, Ag, and Cd, respectively. Row labels 1, 2, and 3 represent the first principal components through third, and the rows show the total variation due to each successive element included in the subset.

| | **123** | **124** | **125** | **126** | **127** | **134** | **135** | **136** | **137** |
|---|---|---|---|---|---|---|---|---|---|
| **1** | 76.892 | 26.838 | 27.477 | 26.829 | 28.109 | 73.801 | 73.957 | 73.786 | 74.254 |
| **2** | 96.404 | 35.383 | 36.032 | 35.373 | 53.809 | 86.312 | 86.730 | 86.294 | 100.820 |
| **3** | 104.564 | 37.340 | 38.204 | 35.879 | 62.344 | 88.269 | 89.133 | 86.808 | 113.274 |

| | **145** | **146** | **147** | **156** | **157** | **167** | **234** | **235** | **236** |
|---|---|---|---|---|---|---|---|---|---|
| **1** | 17.553 | 17.110 | 27.027 | 17.534 | 27.071 | 27.027 | 71.675 | 71.838 | 71.661 |
| **2** | 20.218 | 19.223 | 44.089 | 19.991 | 44.535 | 44.074 | 87.537 | 88.137 | 87.529 |
| **3** | 21.909 | 19.584 | 46.049 | 20.448 | 46.914 | 44.589 | 89.498 | 90.362 | 88.037 |

| | **237** | **245** | **246** | **247** | **256** | **257** | **267** | **345** | **346** |
|---|---|---|---|---|---|---|---|---|---|
| **1** | 72.186 | 18.941 | 18.335 | 27.146 | 18.938 | 27.216 | 27.146 | 69.371 | 69.243 |
| **2** | 98.651 | 21.493 | 20.457 | 45.309 | 21.220 | 45.926 | 45.308 | 72.353 | 71.377 |
| **3** | 114.503 | 23.138 | 20.813 | 47.278 | 21.677 | 48.143 | 45.818 | 74.067 | 71.742 |

| | **347** | **356** | **357** | **367** | **456** | **457** | **467** | **567** |
|---|---|---|---|---|---|---|---|---|
| **1** | 69.771 | 69.357 | 69.891 | 69.758 | 3.272 | 27.030 | 26.998 | 27.030 |
| **2** | 96.234 | 72.149 | 96.367 | 96.221 | 5.039 | 30.136 | 29.156 | 29.929 |
| **3** | 98.208 | 72.606 | 99.072 | 96.747 | 5.382 | 31.847 | 29.522 | 30.387 |

TABLE H.2 Total Variance (Compare with 136.944 Total Variance) for Three-Component Subsets, in Order of Increasing Variance.

| **456** | **146** | **156** | **246** | **256** | **145** | **245** | **467** | **567** | **457** |
|---|---|---|---|---|---|---|---|---|---|
| 5.382 | 19.584 | 20.448 | 20.813 | 21.677 | 21.909 | 23.138 | 29.522 | 30.387 | 31.847 |
| **126** | **124** | **125** | **167** | **267** | **147** | **157** | **247** | **257** | **127** |
| 35.879 | 37.340 | 38.204 | 44.589 | 45.818 | 46.049 | 46.914 | 47.278 | 48.143 | 62.344 |
| **346** | **356** | **345** | **136** | **236** | **134** | **135** | **234** | **235** | **367** |
| 71.742 | 72.606 | 74.067 | 86.808 | 88.037 | 88.269 | 89.133 | 89.498 | 90.362 | 96.747 |
| **347** | **357** | **123** | **137** | **237** | | | | | |
| 98.208 | 99.072 | 104.564 | 113.274 | 114.503 | | | | | |

TABLE H.3 Principal Components Analysis on All Four-Element Subsets of 1,373-Bullet Subset. Elements 1, 2, 3, 4, 5, 6, and 7 are As, Sb, Sn, Cu, Bi, Ag, and Cd, respectively. Row labels 1, 2, 3, and 4 represent the first principal component through fourth, and the rows show the total variation due to each successive element included in the subset.

| | **1234** | **1235** | **1236** | **1237** | **1245** | **1246** | **1247** | **1256** | **1257** |
|---|---|---|---|---|---|---|---|---|---|
| **1** | 76.918 | 77.133 | 76.903 | 77.362 | 27.517 | 26.865 | 28.126 | 27.506 | 28.599 |
| **2** | 96.441 | 97.085 | 96.430 | 103.955 | 36.072 | 35.410 | 53.844 | 36.061 | 54.501 |
| **3** | 104.603 | 105.249 | 104.590 | 123.430 | 38.556 | 37.516 | 62.380 | 38.279 | 63.047 |
| **4** | 106.557 | 107.421 | 105.096 | 131.562 | 40.197 | 37.872 | 64.337 | 38.736 | 65.202 |

| | **1267** | **1345** | **1346** | **1347** | **1356** | **1357** | **1367** | **1456** | **1457** |
|---|---|---|---|---|---|---|---|---|---|
| **1** | 28.122 | 73.982 | 73.810 | 74.278 | 73.966 | 74.440 | 74.263 | 17.575 | 27.071 |
| **2** | 53.835 | 86.772 | 86.330 | 100.843 | 86.751 | 101.012 | 100.828 | 20.366 | 44.575 |
| **3** | 62.371 | 89.436 | 88.440 | 113.309 | 89.208 | 113.752 | 113.291 | 22.099 | 47.221 |
| **4** | 62.877 | 91.126 | 88.801 | 115.267 | 89.665 | 116.131 | 113.806 | 22.441 | 48.906 |

| | **1467** | **1567** | **2345** | **2346** | **2347** | **2356** | **2357** | **2367** | **2456** |
|---|---|---|---|---|---|---|---|---|---|
| **1** | 27.027 | 27.071 | 71.861 | 71. 683 | 72.209 | 71.847 | 72.378 | 72.195 | 18.969 |
| **2** | 44.108 | 44.556 | 88.174 | 87.562 | 98.673 | 88.164 | 98.855 | 98.660 | 21.650 |
| **3** | 46.221 | 46.989 | 90.710 | 89.674 | 114.534 | 90.437 | 115.149 | 114.526 | 23.328 |
| **4** | 46.581 | 47.446 | 92.355 | 90.030 | 116.495 | 90.894 | 117.360 | 115.035 | 23.670 |

| | **2457** | **2467** | **2567** | **3456** | **3457** | **3467** | **3567** | **4567** |
|---|---|---|---|---|---|---|---|---|
| **1** | 27.217 | 27.146 | 27.217 | 69.378 | 69.911 | 69.777 | 69.898 | 27.031 |
| **2** | 45.955 | 45.333 | 45.952 | 72.492 | 96.387 | 96.241 | 96.374 | 30.276 |
| **3** | 48.496 | 47.454 | 48.218 | 74.257 | 99.355 | 98.375 | 99.147 | 32.037 |
| **4** | 50.135 | 47.810 | 48.675 | 74.599 | 101.065 | 98.740 | 99.604 | 32.380 |

TABLE H.4 Total Variance (Compare with 136.944 Total Variance) for Four-Component Subsets, in Order of Increasing Variance.

| **1456** | **2456** | **4567** | **1246** | **1256** | **1245** | **1467** | **1567** | **2467** | **2567** |
|---|---|---|---|---|---|---|---|---|---|
| 22.441 | 23.670 | 32.380 | 37.872 | 38.736 | 40.197 | 46.581 | 47.446 | 47.810 | 48.675 |
| **1457** | **2457** | **1267** | **1247** | **1257** | **3456** | **1346** | **1356** | **2346** | **2356** |
| 48.906 | 50.135 | 62.877 | 64.337 | 65.202 | 74.599 | 88.801 | 89.665 | 90.030 | 90.894 |
| **1345** | **2345** | **3467** | **3567** | **3457** | **1236** | **1234** | **1235** | **1367** | **2367** |
| 91.126 | 92.355 | 98.740 | 99.604 | 101.065 | 105.096 | 106.557 | 107.421 | 113.806 | 115.035 |
| **1347** | **1357** | **2347** | **2357** | **1237** | | | | | |
| 115.267 | 116.131 | 116.495 | 117.360 | 131.562 | | | | | |

TABLE H.5 Principal Components Analysis on All Five-Element Subsets of 1,373-Bullet Subset. Elements 1, 2, 3, 4, 5, 6, and 7 are As, Sb, Sn, Cu, Bi, Ag, and Cd, respectively. Row labels 1, 2, 3, 4, and 5 represent the first principal components through fifth, and the rows show the total variation due to each successive element included in the subset.

| | **12345** | **12346** | **12347** | **12356** | **12357** | **12367** | **12456** | **12457** | **12467** |
|---|---|---|---|---|---|---|---|---|---|
| **1** | 77160 | 76.930 | 77.388 | 77.144 | 77.608 | 77.373 | 27.547 | 28.624 | 28.140 |
| **2** | 97.127 | 96.468 | 103.981 | 97.114 | 104.205 | 103.966 | 36.103 | 54.541 | 53.871 |
| **3** | 105.292 | 104.630 | 123.467 | 105.278 | 124.130 | 123.456 | 38.716 | 63.088 | 62.408 |
| **4** | 107.775 | 106.733 | 131.600 | 107.496 | 132.265 | 131.588 | 40.387 | 65.560 | 64.514 |
| **5** | 109.414 | 107.089 | 133.554 | 107.953 | 134.419 | 132.094 | 40.729 | 67.194 | 64.869 |

| | **12567** | **13456** | **13457** | **13467** | **13567** | **14567** | **23456** | **23457** | **23467** |
|---|---|---|---|---|---|---|---|---|---|
| **1** | 28.617 | 73.991 | 74.464 | 74.286 | 74.448 | 27.072 | 71.870 | 72.401 | 72.217 |
| **2** | 54.530 | 86.795 | 101.037 | 100.852 | 101.021 | 44.598 | 88.203 | 98.878 | 98.682 |
| **3** | 63.076 | 89.584 | 113.794 | 113.328 | 113.773 | 47.372 | 90.867 | 115.186 | 114.559 |
| **4** | 65.277 | 91.316 | 116.440 | 115.438 | 116.206 | 49.096 | 92.546 | 117.714 | 116.617 |
| **5** | 65.734 | 91.658 | 118.124 | 115.799 | 116.663 | 49.439 | 92.887 | 119.353 | 117.028 |

| | **23567** | **24567** | **34567** |
|---|---|---|---|
| **1** | 72.387 | 27.218 | 69.918 |
| **2** | 98.864 | 45.984 | 96.394 |
| **3** | 115.177 | 48.655 | 99.495 |
| **4** | 117.435 | 50.326 | 101.254 |
| **5** | 117.892 | 50.667 | 101.597 |

TABLE H.6 Total Variance (Compare with 136.944 Total Variance) for Five-Component Subsets, in Order of Increasing Variance.

| **12456** | **14567** | **24567** | **12467** | **12567** | **12457** | **13456** | **23456** | **34567** | **12346** |
|---|---|---|---|---|---|---|---|---|---|
| 40.73 | 49.44 | 50.67 | 64.87 | 65.73 | 67.19 | 91.66 | 92.89 | 101.60 | 107.09 |
| %29.74 | 36.10 | 37.00 | 47.37 | 48.00 | 49.07 | 66.93 | 67.83 | 74.19 | 78.20 |

| **12356** | **12345** | **13467** | **12567** | **23467** | **23567** | **13457** | **23457** | **12367** | **12347** |
|---|---|---|---|---|---|---|---|---|---|
| 107.95 | 109.41 | 115.80 | 116.66 | 117.03 | 117.89 | 118.12 | 119.35 | 132.09 | 133.55 |
| 78.83 | 79.90 | 84.56 | 85.19 | 85.46 | 86.09 | 8 6.26 | 87.15 | 96.46 | 97.53 |

| **12357** |
|---|
| 134.42 |
| 98.16 |

TABLE H.7 Principal Components Analysis on All Six-Element Subsets of 1,373-Bullet Subset. Elements 1, 2, 3, 4, 5, 6, and 7 are As, Sb, Sn, Cu, Bi, Ag, and Cd, respectively. Row labels 1, 2, 3, 4, 5, and 6 represent the first principal component through sixth, and the rows show the total variation due to each successive element included in the subset.

| | **123456** | **123457** | **123467** | **123567** | **124567** | **134567** | **234567** |
|---|---|---|---|---|---|---|---|
| **1** | 77.172 | 77.635 | 77.399 | 77.620 | 28.643 | 74.472 | 72.411 |
| **2** | 97.157 | 104.232 | 103.993 | 104.216 | 54.571 | 101.046 | 98.887 |
| **3** | 105.322 | 124.172 | 123.494 | 124.159 | 63.118 | 113.817 | 115.215 |
| **4** | 107.934 | 132.307 | 131.628 | 132.294 | 65.721 | 116.590 | 117.872 |
| **5** | 109.605 | 134.779 | 133.731 | 134.494 | 67.385 | 118.314 | 119.543 |
| **6** | 109.946 | 136.411 | 134.087 | 134.951 | 67.726 | 118.656 | 119.885 |

TABLE H.8 Total Variance (Compare with 136.944 Total Variance) for Six-Component Subsets, in Order of Increasing Variance

| **124567** | **123456** | **134567** | **234567** | **123467** | **123567** | **123457** |
|---|---|---|---|---|---|---|
| 67.726 | 109.946 | 118.656 | 119.885 | 134.087 | 134.951 | 136.411 |
| 49.45% | 80.28% | 86.65% | 87.54% | 97.91% | 98.54% | 99.61% |

TABLE H.9 Principal Components Analysis on all Seven-Element Subsets of 1,373-Bullet Subset. Elements 1, 2, 3, 4, 5, 6, and 7 are As, Sb, Sn, Cu, Bi, Ag, and Cd, respectively. Row labels 1, 2, 3, 4, 5, and 6 represent the first principal component through sixth , and the rows show the total variation due to each successive element included in the subset.

| | **1234567** |
|---|---|
| **1** | 77.64703 |
| **2** | 104.24395 |
| **3** | 124.20241 |
| **4** | 132.33795 |
| **5** | 134.94053 |
| **6** | 136.60234 |
| **7** | 136.94360 |

Summary:

| | | |
|---|---|---|
| 3 elements: | 237 | (83.6% of total variance) |
| 4 elements: | 1237 | (96.07% of total variance) |
| 5 elements: | 12357 | (98.16% of total variance) or 12347 (97.52%) |
| 6 elements: | 123567 | (99.61% of total variance) or 123457 (98.54%) (Bi-Ag correlation) |
| 7 elements: | 1234567 | (100.00% of total variance) |

## REFERENCES

1. Koons, R. D. and Grant, D. M. *J. Foren.. Sci.* **2002**, 47(5), 950.
2. Randich, E.; Duerfeldt, W.; McLendon, W.; and Tobin, W. *Foren. Sci. Int.* **2002**, 127, 174–191.
3. Peele, E. R.; Havekost, D. G.; Peters, C. A.; and Riley, J. P. USDOJ (ISBN 0-932115-12-8), 57, **1991**.

# I

# Birthday Problem Analogy

The committee has found a perceived similarity between determining the false match probability for bullet matches and a familiar problem in probability, the Birthday Problem: Given $n$ people (bullets) in a room (collection), what is the probability that at least two of them share the same birthday (analytically indistinguishable composition)? Ignoring leap-year birthdays (February 29), the solution is obtained by calculating the probability of the complementary event ("P{A}" denotes the probability of the event A):

P{no 2 people have the same birthday} = P{each of $n$ persons has a different birthday} = P{person 1 has any of 365 birthdays} · P{person 2 has any of the other 364 birthdays} · . . . $= \frac{365}{365}\,\frac{364}{365} \cdots \frac{365-n+1}{365} = \bar{p}(n)$.

Then P{at least 2 people have the same birthday} $= p(n) = 1 - \bar{p}(n)$. When $n = 6$, 23, 55, $p(n) = 0.04$, 0.51, 0.99, respectively (Ref. 1).

That calculation seems to suggest that the false match probability is extremely high when the case contains 23 or more bullets, but the compositional analysis of bullet lead (CABL) matching problem differs in three important ways.

- First, CABL attempts to match not just *any* two bullets (which is what the birthday problem calculates), but one *specific* crime scene bullet and one or more of $n$ other potential suspect bullets where $n$ could be as small as 1 or 2 or as large as 40 or 50 (which is similar to determining the probability that a specific person shares a birthday with another person in the group). Hence, bullet matching by CABL is a completely different calculation from the birthday problem.

• Second, as stated in Chapter 4, a match indicates that the two bullets probably came from the same source or from compositionally indistinguishable volumes of lead, of which thousands exist from different periods, for different types of bullets, for different manufacturers, and so on, not just 365. Even if interest lay in the probability of a match between *any* two bullets, the above calculation with $N = 5{,}000$ and $n + 1 = 6$ or 23 or 55 bullets yields much smaller probabilities of 0.003, 0.0494, and 0.2578, respectively.

• Third, if bullets manufactured at the same time tend to appear in the same box, and such boxes tend to be distributed in geographically nondispersed locations, the $n$ potential suspect bullets are not independent, as the $n$ persons' birthdays in the birthday problem are assumed to be.

We conclude that this analogy with the birthday problem does not apply.

## REFERENCE

1. Chung, K.-L. *Elementary Probability Theory with Stochastic Processes, 2nd Ed.*, Springer-Verlag: New York, NY, 1975; p 63.

# J

# Understanding the Significance of the Results of Compositional Analysis of Bullet Lead

As explained in Chapter 4, there is a need for the meaning of technical evidence to be elucidated for use by attorneys, judges, and jury members. This appendix is intended to serve as a rough guideline for such information to be included in a "boiler plate" document. Such a document would be attached to or incorporated in laboratory reports dealing with compositional analysis of bullet lead (CABL). It is not necessarily intended to be used as is.

## INTRODUCTION

CABL can provide useful and probative information regarding a possible association between known bullets and questioned bullets or bullet fragments in a number of case situations. However, CABL has its limits, and the strength of an indicated association will vary from case to case. Care should be taken that these limitations and caveats are appreciated and understood; this may require expert interpretation.

## CHEMICAL ANALYSIS

CABL uses a chemical technique called inductively coupled plasma-optical emission spectroscopy (ICP-OES). It is capable of detecting and measuring the concentrations of several elements that occur as trace impurities in or minor alloying elements of bullet lead. The concentrations of those elements that are most useful in discriminating among bullet leads can be measured with good sensitivity, accuracy, and precision. Close correspondence of the quantitative measurements between two samples (the samples are "analytically indistinguishable") may suggest that the two samples were derived from a common "source." However, several poorly characterized processes in the production of bullet lead and ammunition, as well as in ammunition distribution, complicate the interpretation and render a definition of "source" difficult. For that reason, unlike the situation with some forms of evidence (such as the DNA typing of bloodstains), it is not possible to obtain accurate and easily understood probability estimates that are directly applicable. It is necessary for the finder of fact to have a general understanding of the possible complicating factors.

## OVERVIEW OF THE GENERAL MANUFACTURING PROCESS

Virtually all the lead used in the manufacture of lead bullets and lead bullet cores in the United States is purchased from secondary lead smelters that use recycled automotive batteries as their primary source of lead. It is not economically feasible to attempt to remove particular elements below some point. To meet user specifications during the refining process, smelters must keep the concentrations of specified elements in the lead within a range or below a maximum set by the bullet manufacturers. The variation in several elements from the ore, from use as battery lead, and required by the bullet manufacturers (arsenic, As; antimony, Sb; tin, Sn; copper, Cu; bismuth, Bi; silver, Ag; and cadmium, Cd) provides the basis of discrimination used in CABL. The smelter casts the refined molten lead into molds, where it cools and solidifies to form castings for shipment to customers, including bullet manufacturers. A variety of mold sizes can be used to produce castings known as *pigs*, *sows*, *ingots*, and *billets*.

Bullet manufacturers produce bullets from continuous cylindrical *wires* of lead. The wires are produced by extrusion, when the billet is forced through a circular orifice of a specified size to produce the lead wire. The diameter of the wire produced depends on the caliber and design of the bullets to be made. Some bullet-manufacturing plants obtain billets for wire extrusion directly from the smelter. Others produce their own billets from large melts made from larger castings obtained from the lead smelter.

Additional steps in the bullet-manufacturing process can introduce changes in the lead's elemental composition. When ingots are melted in the bullet-manufacturing plant, multiple ingots of different composition may be melted together in a large vessel. In addition, the composition of the melt may change because of oxidation of some elements by exposure to air, the addition of lead recycled from other parts of the operation, and drawing off of molten lead for casting while lead is being added to the vessel. Thus, small but important changes in the composition of the lead can take place during many steps in the smelting and bullet-production steps.

Furthermore, as a billet cools, any radial segregation that occurs tends to be homogenized during extrusion of the wire. Top-to-bottom variations still exist, but it is probable that the industry practice of removing the first several feet of extruded wire will remove much of the wire that has noticeably different lead characteristics. After this point, the lead will maintain the same composition indefinitely.

## ASSEMBLY OF AMMUNITION

The extruded wire is cut into segments to form slugs that will become bullets and bullet cores; these may be stockpiled in bins, possibly with slugs from different wires with different compositions, before they are assembled with other components to form cartridges. Bullets from multiple bins (also with different compositions) may be assembled into cartridges at the same time. That results in the possibility that different compositions of bullet lead are present in a single box of ammunition.

## AMMUNITION DISTRIBUTION

Details about the manufacturing and distribution of lead bullets and finished ammunition are largely unavailable. Therefore, distribution patterns and their effect on random matches cannot be estimated. Calculations can be used only to offer general guidance in assessing the significance of a finding that certain bullets are analytically indistinguishable.

## DEFINITION OF SOURCE

The previously mentioned uncertainties arising from factors related to manufacturing make it difficult to define the size of a "source," hereafter referred to as a *compositionally indistinguishable volume of lead* (CIVL). The analytically indistinguishable regions of wire could be considered a CIVL, but other wires extruded from billets from the same melt (assuming there was no additional material added to the melt while the lead was being poured) could also have regions that are analytically indistinguishable from this first wire (although this has not been confirmed by a quantitative, scientific study). A CIVL may range from approximately 70 lbs in a billet to 200,000 lbs in a melt. That is equivalent to 12,000 to 35 million .22 caliber bullets in a CIVL out of a total of 9 billion bullets produced each year.

## RANDOM COINCIDENTAL MATCHES

Although it would be extraordinarily difficult—or impossible—for a large-scale industrial operation (smelter or bullet manufacturer) to purposefully duplicate a given CIVL, the possibility of recurrence of a composition over time as an occasional random event cannot be dismissed. Theoretically, the number of these that might repeat would depend in part on the number of elements measured, the permitted concentration range for each element, and the discrimination of the analytical technique for each. Considering the thousands of "batches" of lead produced over a number of years, there is a reasonably high probability that some will repeat. However, the probability that any given composition would repeat within the next several years could be expected to be quite low. Furthermore, the likelihood that such a coincidental match would occur from such a source and appear in a given case would be smaller still.

In summary, a CIVL would be large in comparison with the amount of lead in the ammunition in the possession of a typical single purchaser. Thus, multiple people would be expected to have ammunition with the same lead composition. It is not known how many of these would be in the same geographic area. As time passes and some of the ammunition is used, the likelihood of a false association because of the distribution of ammunition with lead from the same CIVL would decrease.

## MULTIPLE COMPOSITIONS IN A SINGLE CASE

If several evidence bullets in a case have similar but distinguishable compositions, and each of these compositions has a counterpart in a known source, such as a box of ammunition, the association would be stronger.

# K

# Statistical Analysis of Bullet Lead Data

By Karen Kafadar and Clifford Spiegelman

## 1. INTRODUCTION

The current procedure for assessing a "match" (analytically indistinguishable chemical compositions) between a crime-scene (CS) bullet and a potential suspect's (PS) bullet starts with three pieces from each bullet or bullet fragment. Nominally each piece is measured in triplicate with inductively coupled plasma–optical emission spectrophotometry (ICP-OES) on seven elements: As, Sb, Sn, Cu, Bi, Ag, Cd, against three standards. Analyses in previous years measured three to six elements; in some cases, fewer than three pieces can be abstracted from a bullet or bullet fragment. Parts of the analysis below will consider fewer than seven elements, but we will always assume measurements on three pieces in triplicate even though occasionally very small bullet fragments may not have yielded three measurements. The three replicates on each piece are averaged, and then means, standard deviations (SDs), and ranges (minimum to maximum) for the three pieces and for each element are calculated for all CS and PS bullets. Throughout this appendix, the three averages (from the triplicate readings) on the three pieces are denoted the three "measurements" (even though occasionally very small bullet fragments may not have yielded three measurements).

Once the chemical analysis has been completed, a decision must be based on the measurements. Are the data consistent with the hypothesis that the mean chemical concentrations of the two bullets are the same or different? If the data suggest that the mean chemical concentrations are the same, the bullets or fragments are assessed as "analytically indistinguishable." Intuitively, it makes sense that if the seven average concentrations (over the three measurements) of the CS bullet are "far" from those of the PS bullet, the data would be deemed more consistent with the hypothesis of "no match." But if the seven averages are "close," the data would be more consistent with the hypothesis that the two bullets "match." The role of statistics is to determine *how close*, that is, to determine limits beyond which the bullets are deemed to have come from sources that have different mean concentrations and within which they are deemed to have come from sources that have the same mean concentrations.

### 1.1. Statistical Hypothesis Tests

The classical approach to deciding between the two hypotheses was developed in the 1930s. The standard hypothesis-testing procedure consists of these steps:

1. Set up the two hypotheses. The "assumed" state of affairs is generally the *null hypothesis*, for example, "drug is no better than placebo." In the compositional analysis of bullet lead (CABL) context, the null hypothesis is "bullets do not match" or "mean concentrations of materials from which these two bullets were produced are not the same" (assume "not guilty"). The converse is called the *alternative hypothesis*, for example, "drug is effective" or in the CABL context, "bullets match" or "mean concentrations are the same."
2. Determine an *acceptable level of risk* posed by rejecting the null hypothesis when it is actually true. The level is set according to the circumstances. Conventional values in many fields are 0.05 and 0.01; that is, in one of 20 or in one of 100 cases when this test is conducted, the test will erroneously decide on the alternative hypothesis ("bullets match") when the null hypothesis actually was correct ("bullets do not match"). The preset level is considered inviolate; a procedure will not be considered if its "risk" exceeds it. We consider below tests with desired risk levels of 0.30 to 0.0004. (The value of 0.0004 is equivalent to 1 in 2,500, thought by the FBI to be the current level.)
3. Calculate a quantity based on the data (for example, involving the sample mean concentrations of the seven elements in the two bullets), known as a *test statistic*. The value of the test statistic will be used to test the null hypothesis versus the alternative hypothesis.
4. The preset level of risk and the test statistic together define two regions, corresponding to the two hypotheses. If the test statistic falls in one region, the decision is to fail to reject the null hypothesis; if it falls in the other region (called the *critical region*), the decision is to reject the null hypothesis and conclude the alternative hypothesis.

The critical region has the following property: Over the many times that this protocol is followed, the probability of falsely rejecting the null hypothesis does not exceed the preset level of risk. The recommended test procedure in Section 4 has a further property: if the alternative hypothesis holds, the procedure will have the greatest chance of correctly rejecting the null hypothesis.

The FBI protocol worked in reverse. Three test procedures were proposed, described below as "2-SD overlap," "range overlap," and "chaining." Thus, the first task of the authors was to calculate the level of risk that would result from the use of these three procedures. More precisely, we developed a simulation, guided by information about the bullet concentrations from various sources and from datasets that were published or provided to the committee (described in Section 3.2), to calculate the probability that the 2-SD-overlap and range-overlap procedures would claim a match between two bullets whose mean concentrations differed by a specified amount. The details of that simulation and the resulting calculations are described in Section 3.3 with a discussion of chaining.

An alternative approach, based on the theory of equivalence *t* tests, is presented in Section 4. A level of risk is set for each equivalence *t* test to compare two bullets on each of the seven elemental concentrations; if the mean concentrations of all seven elements are sufficiently close, the overall false-positive probability (FPP) of a match between two bullets that actually differ is less than 0.0004 (one in 2,500). The method is described in detail so that the reader can apply it with another value of the FPP such as one in 500, or one in 10,000. A multivariate version of the seven separate tests (Hotelling's $T^2$) is also described. Details of the statistical theory are provided in the other appendixes. Appendix E contains basic principles of statistics; Appendix F provides a theoretical derivation that characterizes the FBI procedures and equivalence tests and some extra analyses not shown in this appendix; Appendix H describes the principal-component analysis for assessing the added contributions of each element for purposes of discrimination; and Appendix G provides further analyses conducted on the data sets.

## 1.2 Current Match Procedure

The FBI presented three procedures for assessing a match between two bullets:

- "2-SD overlap." Measurements of each element can be combined to form an interval with lower limit *mean –2SD* and upper limit *mean+2SD*. The means and SDs are based on the average of three measurements in each of the specimens. If the seven intervals for a given CS bullet overlap with all seven intervals for a given PS bullet, the CS and PS bullets are deemed a match.
- "Range overlap." Intervals for each element are calculated as minimum to maximum from the three measurements in each of the specimens. If the seven intervals for a given CS bullet overlap with all seven intervals for a given PS bullet, the CS and PS bullets are deemed a match.

• Chaining. As described in FBI Laboratory document *Comparative Elemental Analysis of Firearms Projectile lead by ICP-OES* (Ref. 1, pp. 10–11):

> a. CHARACTERIZATION OF THE CHEMICAL ELEMENT DISTRIBUTION IN THE KNOWN PROJECTILE LEAD POPULATION
>
> The mean element concentrations of the first and second specimens in the known material population are compared based upon twice the measurement uncertainties from their replicate analysis. If the uncertainties overlap in all elements, they are placed into a composition group; otherwise they are placed into separate groups. The next specimen is then compared to the first two specimens, and so on, in the same manner until all of the specimens in the known population are placed into compositional groups. Each specimen within a group is analytically indistinguishable for all significant elements measured from at least one other specimen in the group and is distinguishable in one or more elements from all the specimens in any other compositional group. (It should be noted that occasionally in groups containing more than two specimens, chaining occurs. That is, two specimens may be slightly separated from each other, but analytically indistinguishable from a third specimen, resulting in all three being included in the same compositional group.)
>
> b. COMPARISON OF UNKNOWN SPECIMEN COMPOSITION(S) WITH THE COMPOSITION(S) OF THE KNOWN POPULATION(S)
>
> The mean element concentrations of each individual questioned specimen are compared with the element concentration distribution of each known population composition group. The concentration distribution is based on the mean element concentrations and twice the standard deviation of the results for the known population composition group. If all mean element concentrations of a questioned specimen overlap within the element concentration distribution of one of the known material population groups, that questioned specimen is described as being "analytically indistinguishable" from that particular known group population.

The SD of the "concentration distribution" is calculated as the SD of the averages (over three measurements for each bullet) from all bullets in the "known population composition group." In Ref. 2, the authors (Peele et al. 1991) apply this "chaining algorithm" on intervals formed by the ranges (minimum and maximum of three measurements) rather than (mean ± 2SD) intervals.

The "2-SD overlap" and "range-overlap" procedures are illustrated with data from an FBI-designed study of elemental concentrations of bullets from different boxes (Ref. 2). The three measurements in each of three pieces of each of seven elements (in units of parts per million, ppm) are shown in Table K.1 below for bullets F001 and F002 from one of the boxes of bullets provided by Federal Cartridge Company (described in more detail in Section 3.2). Each piece was mea-

TABLE K.1 Illustration of Calculations for 2-SD-Overlap and Range-Overlap Methods on Federal Bullets F001 and F002 (Concentrations in ppm)

| | icpSb | icpCu | icpAg | icpBi | icpAs | icpSn |
|---|---|---|---|---|---|---|
| | Federal Bullet F001 | | | | | |
| a | 29276 | 285 | 64 | 16 | 1415 | 1842 |
| b | 29506 | 275 | 74 | 16 | 1480 | 1838 |
| c | 29000 | 283 | 66 | 16 | 1404 | 1790 |
| mean | 29260.67 | 281.00 | 68.00 | 16 | 1433.00 | 1823.33 |
| SD | 253.35 | 5.29 | 5.29 | 0 | 41.07 | 28.94 |
| Mean – 2SD | 28753.97 | 270.42 | 57.42 | 16 | 1350.85 | 1765.46 |
| Mean + 2SD | 29767.36 | 291.58 | 78.58 | 16 | 1515.15 | 1881.21 |
| minimum | 29000 | 275 | 64 | 16 | 1404 | 1790 |
| maximum | 29506 | 285 | 74 | 16 | 1480 | 1842 |
| | Federal Bullet F002 | | | | | |
| | icpSb | icpCu | icpAg | icpBi | icpAs | icpSn |
| a | 28996 | 278 | 76 | 16 | 1473 | 1863 |
| b | 28833 | 279 | 67 | 16 | 1439 | 1797 |
| c | 28893 | 282 | 77 | 15 | 1451 | 1768 |
| mean | 28907.33 | 279.67 | 73.33 | 15.67 | 1454.33 | 1809.33 |
| SD | 82.44 | 2.08 | 5.51 | 0.58 | 17.24 | 48.69 |
| mean – 2SD | 28742.45 | 275.50 | 62.32 | 14.51 | 1419.84 | 1711.96 |
| mean + 2SD | 29072.21 | 283.83 | 84.35 | 16.82 | 1488.82 | 1906.71 |
| minimum | 28833 | 278 | 67 | 15 | 1439 | 1768 |
| maximum | 28996 | 282 | 77 | 16 | 1473 | 1863 |

sured three times against three different standards; only the average is provided, and in this report it is called the "measurement." Table K.1 shows the three measurements, their means, their SDs (equal to the square root of the sum of the three squared deviations from the mean divided by 2), the "2-SD interval" (mean –2SD to mean + 2SD), and the "range interval" (minimum and maximum).

For all seven elements, the 2-SD interval for Federal bullet 1 overlaps with the 2-SD interval for Federal bullet 2. Equivalently, the difference between the means is less than twice the sum of the two SDs. For example, the 2-SD interval for Cu in bullet 1 is (270.42, 291.58), and the interval for Cu in bullet 2 is (275.50, 283.83), which is completely within the Cu 2-SD interval for bullet 1. Equivalently, the difference between the means (281.00 and 279.67) is 1.33, less than 2(5.29 + 2.08) is 14.74. Thus, the 2-SD overlap procedure would conclude that the two bullets are analytically indistinguishable (Ref. 3) on all seven elements, so the bullets would be claimed to be analytically indistinguishable. The range overlap procedure would find the two bullets analytically indistinguishable on all elements except Sb because for all other elements the range interval on each element for bullet 1 overlaps with the corresponding interval for bullet 2; for example, for Cu (275, 285) overlaps with (278, 282), but for Sb, the range interval (29,000, 29,506) just fails to overlap (28,833, 28,996) by only 4 ppm. Hence, by the range-overlap procedure, the bullets would be analytically distinguishable.

## 2. DESCRIPTION AND ANALYSIS OF DATASETS

### 2.1 Description of Data Sets

This section describes three data sets made available to the authors in time for analysis. The analysis of these data sets resulted in the following observations:

1. The uncertainty in measuring the seven elements is usually 2.0–5.0%.
2. The distribution of the measurements is approximately lognormally distributed; that is, logarithms of measurements are approximately normally distributed. Because the uncertainty in the three measurements on a bullet is small (frequently less than 5%), the lognormal distribution with a small relative SD is similar to a normal distribution. For purposes of comparing the measurements on two bullets, the measurements need not be transformed with logarithms, but it is often more useful to do so.
3. The distributions of the concentrations of a given element across many different bullets from various sources are lognormally distributed with much more variability than seen from within-bullet measurement error or within-lot error. For purposes of comparing average concentrations across many different bullets, the concentrations should be transformed with logarithms first, and then means and SDs can be calculated. The results can be reported on the original scale by taking the antilogarithms for example, exp(mean of logs).
4. The errors in the measurements of the seven elements may not be uncorrelated. In particular, the errors in measuring Sb and Cu appear to be highly correlated (correlation approximately 0.7); the correlation between the errors in measuring Ag and Sb or between the errors in measuring Ag and Cu is approximately 0.3. Thus, if the 2-SD intervals for Sb on two bullets overlap, the 2-SD intervals for Cu may be more likely to overlap also.

These observations will be described during the analysis part of this section.

The three data sets that were studied by the authors are denoted here as "800-bullet data set," "1,837-bullet data set," and "Randich et al. data set."

1. *800-bullet data set* (Ref. 4): This data set contains triplicate measurements on 50 bullets in each of four boxes from each of four manufacturers—

CCI, Federal, Remington, and Winchester—measured as part of a careful study conducted by Peele et al. (1991). Measured elements in the bullet lead were Sb, Cu, and As, measured with neutron activation analysis (NAA), and Sb, Cu, Bi, and Ag (measured with ICP-OES). In the Federal bullet lead, As and Sn were measured with NAA and ICP-OES. This 800-bullet data set provided individual measurements on the three bullet lead samples which permitted calculation of means and SDs on the log scale and within-bullet correlations among six of the seven elements measured with ICP-OES (As, Sb, Sn, Bi, Cu, and Ag); see Section 3.2.

2. *1,837-bullet data set* (Ref. 5): The bullets in this data set were extracted from a larger, historical file of 71,000+ bullets analyzed by the FBI Laboratory during the last 15 years. According to the notes that accompanied the data file, the bullets in it were selected to include one bullet (or sometimes more) that were determined to be distinct from the other bullets in the case; a few are research samples "not associated with any particular case," and a few "were taken from the ammunition collection (again, not associated with a particular case)." The notes that accompanied this data set stated:

> To assure independence of samples, the number of samples in the full data set was reduced by removing multiple bullets from a given known source in each case. To do this, evidentiary submissions were considered one case at a time. For each case, one specimen from each combination of bullet caliber, style, and nominal alloy class was selected and that data was placed into the test sample set. In instances where two or more bullets in a case had the same nominal alloy class, one sample was randomly selected from those containing the maximum number of elements measured. . . . The test set in this study, therefore, should represent an unbiased sample in the sense that each known production source of lead is represented by only one randomly selected specimen. [Ref. 6]

All bullets in this subset were measured three times (three fragments). Bullets from 1,005 cases between 1989 and 2002 are included; in 528 of these cases, only one bullet was selected. The numbers of cases for which different numbers of bullets were selected are given in Table K.2. The cases that had 11, 14, and 21 bullets were cases 834, 826, and 982, respectively. Due to the way in which these bullets were selected, they do not represent a random sample of bullets from any population—even the population of bullets analyzed by the laboratory. The selection probably produced a data set whose variability among bullets is higher than might be seen in the complete data set or in the population of all manufactured bullets. Only averages and SDs of the (unlogged) measurements are available, not the

TABLE K.2 Number of Cases Having *b* Bullets in the 1,837-Bullet Data Set

| $b$ = no. bullets | 1 | 2 | 3 | 4 | 5 | 6 | 7 | 8 | 9 | 10 | 11 | 14 | 21 |
|---|---|---|---|---|---|---|---|---|---|---|---|---|---|
| No. cases | 578 | 238 | 93 | 48 | 24 | 10 | 7 | 1 | 1 | 2 | 1 | 1 | 1 |

three individual measurements themselves, so a precise estimate of the measurement uncertainty (relative SD within bullets) could not be calculated, as it could in the 800-bullet data set. (One of the aspects of the nonrandomness of this dataset is that it is impossible to determine whether the "selected" bullets tended to have larger or smaller relative SDs (RSDs) compared with the RSDs on all 71,000+ bullets.) Characteristics of this data set are given in Table K.3. Only Sb and Ag were measured in all 1,837 bullets in this data set; all but three of the 980 missing Cd values occurred within the first 1,030 bullets (before 1997). In only 854 of the 1,837 bullets were all seven elements measured; in 522 bullets, six elements were measured (in all but three of the 522 bullets, the missing element is Cd); in 372 bullets, only five elements are measured (in all but 10 bullets, the missing elements are Sn and Cd); in 86 bullets, only four elements are measured (in all but eight bullets, the missing elements are As, Sn, and Cd). The data on Cd are highly discrete: of the 572 nonzero measured averages (139, 96, 40, 48, 32, and 28) showed average Cd concentrations of only (10, 20, 30, 40, 50, and 60) ppm respectively (0.00001–0.00006). The remaining 189 nonzero Cd concentrations were spread out from 70 to 47,880 ppm (0.00007 to 0.04788). This data set provided some information on distributions of averages of the various elements and some correlations between the averages.

Combining the 854 bullets in which all seven elements were measured with the 519 bullets in which all but Cd were measured yielded a subset of 1,373 bullets in which only 519 values of Cd needed to be imputed (estimated from the data). These 1,373 bullets then had measurements on all seven elements. The average Cd concentration in a bullet appeared to be uncorrelated with the average concentration of any other element, so the missing Cd concentration in 519 bullets was imputed by selecting at random one of the 854 Cd values measured in the 854 bullets in which all seven elements were measured. The 854- and 1,373-bullet subsets were used in some of the analyses below.

*3. Randich et al. (2002)* (Ref. 7): These data come from Table 1 of the article by Randich et al. (Ref. 7). Six elements (all but Cd) were measured in three pieces of wire from 28 lots of wire. The three pieces were selected from the beginning, middle, and end of the wire reel. The analysis of this data set confirms the homogeneity of the material in a lot within measurement error.

TABLE K.3 Characteristics of 1,837-Bullet Data Set

| Element | As | Sb | Sn | Bi | Cu | Ag | Cd |
|---|---|---|---|---|---|---|---|
| No. bullets with no data | 87 | 0 | 450 | 8 | 11 | 0 | 980 |
| No. bullets with data | 1,750 | 1,837 | 1,387 | 1,829 | 1,826 | 1,837 | 857 |
| No. bullets with nonzero data | 1,646 | 1,789 | 838 | 1,819 | 1,823 | 1,836 | 572 |
| pooled RSD,% | 2.26 | 2.20 | 2.89 | 0.66 | 1.48 | 0.58 | 1.39 |

## 2.2 Lognormal Distributions

The SDs of measurements made with ICP-OES tend to be proportional to their means; hence, one typically refers to *relative* standard deviation, usually expressed as 100% × (SD/mean). When the measurements are transformed first via logarithms, the SD of the log(measurements) is approximately, and conveniently, equal to the RSD on the original scale. That is, the SD on the log scale will be very close to the RSD on the original scale. The mathematical details of this result are given in Appendix E. A further benefit of the transformation is that the resulting transformed measurements have distributions that are much closer to the familiar normal (Gaussian) distribution—an assumption that underlies many classical statistical procedures. The 800-bullet data set allowed calculation of the RSD by calculating the ordinary SD on the logarithms of the measurements.

The bullet means in the 1,837-bullet data set tend to be lognormally distributed, as shown by the histograms in Figures 3.1–3.4. The data on log(Sn) show two modes, and the data on Sb are split into Sb < 0.05 and Sb > 0.05. The histograms suggest that the concentrations of Sb and Sn in this data set consist of mixtures of lognormal distributions.) Carriquiry et al. (Ref. 8) also used lognormal distributions in analyzing the 800-bullet datas et.

Calculating means and SDs on the log scale was not possible with the data in the 1,837-bullet data set, because only means and SDs of the three measurements are given. However, when the RSD is very small (say, less than 5%), the difference between the lognormal and normal distributions is very small. For about 80% of the bullets in the 1,837-bullet data set that was true for the three measurements of As, Sb, Bi, Cu, and Ag.

## 2.3 Within-Bullet Variances and Covariances

### 800-Bullet Data Set

From the 800-bullet data set, which contains the three measurements in each bullet (not just the mean and SD), one can estimate the measurement SD in each set of three measurements. As mentioned above, when the RSD is small, the lognormally distributed measurement error will have a distribution that is close to normal. The within-bullet covariances shown below were calculated on the log-transformed measurements (results on the untransformed measurements were very similar).

The 800-bullet data set (200 bullets from each of four manufacturers) permits estimates of the within-bullet variances and covariances as follows:

$$s_{jl} = \sum_{k=1}^{200} \sum_{i=1}^{3} \left[ (x_{ijk} - \bar{x}._{jk})(x_{ilk} - \bar{x}._{lk}) / 2 \right] / 200 \quad l, j = 1, \ldots J = \text{number of elements} \quad (1)$$

where $x_{ijk}$ denotes the logarithm of the $i^{th}$ measurement ($i$ = 1 ,2, 3; called "a, b, c" in the data file) of element $j$ in bullet $k$, and $\bar{x}._{jk}$ is the mean of three log(measurements) of element $j$, bullet $k$. When $l = j$, the formula $s_{jj}$ reduces to a pooled within-bullet sample variance for the $j^{th}$ element; compare Equations E.2 and E.3 in Appendix E. Because $s_{jj}$ is based on within-bullet SDs from 200 bullets, the square root of $s_{jj}$ (called a pooled standard deviation) provides a more accurate and precise estimate of the measurement uncertainty than an SD based on only one bullet with three measurements (see Appendix F). The within-bullet

TABLE K.4 Within-Bullet Covariances, times $10^5$, by Manufacturer (800-Bullet Data Set)

| | NAA-As | ICP-Sb | ICP-Cu | ICP-Bi | ICP-Ag |
|---|---|---|---|---|---|
| | | | CCI | | |
| NAA-As | 118 | 10 | 6 | 4 | 17 |
| ICP-Sb | 10 | 48 | 33 | 34 | 36 |
| ICP-Cu | 6 | 33 | 46 | 31 | 36 |
| ICP-Bi | 4 | 34 | 31 | 193 | 29 |
| ICP-Ag | 17 | 36 | 36 | 29 | 54 |
| | | | Federal | | |
| | NAA-As | ICP-Sb | ICP-Cu | ICP-Bi | ICP-Ag |
| NAA-AS | 34 | 8 | 6 | 15 | 7 |
| ICP-Sb | 8 | 37 | 25 | 18 | 39 |
| ICP-Cu | 6 | 25 | 40 | 14 | 42 |
| ICP-Bi | 15 | 18 | 14 | 90 | 44 |
| ICP-Ag | 7 | 39 | 42 | 44 | 681 |
| | | | Remington | | |
| | NAA-As | ICP-Sb | ICP-Cu | ICP-Bi | ICP-Ag |
| NAA- | 345 | –1 | –3 | 13 | 3 |
| ICP-Sb | –1 | 32 | 21 | 16 | 18 |
| ICP-Cu | –3 | 21 | 35 | 15 | 12 |
| ICP-Bi | 13 | 16 | 15 | 169 | 18 |
| ICP-Ag | 3 | 18 | 12 | 18 | 49 |
| | | | Winchester | | |
| | NAA-As | ICP-Sb | ICP-Cu | ICP-Bi | ICP-Ag |
| NAA-As | 555 | 5 | 7 | –5 | 16 |
| ICP-Sb | 5 | 53 | 42 | 45 | 27 |
| ICP-Cu | 7 | 42 | 69 | 37 | 31 |
| ICP-Bi | –5 | 45 | 37 | 278 | 31 |
| ICP-Ag | 16 | 27 | 31 | 31 | 51 |

*continued*

TABLE K.4 *continued*

| | Average over manufacturers | | | | |
|---|---|---|---|---|---|
| | Naa-As | ICP-Sb | ICP-Cu | ICP-Bi | ICP-Ag |
| NAA-As | 263 | 6 | 4 | 7 | 10 |
| ICP-Sb | 6 | 43 | 30 | 28 | 30 |
| ICP-Cu | 4 | 30 | 47 | 24 | 30 |
| ICP-Bi | 7 | 28 | 24 | 183 | 30 |
| ICP-Ag | 10 | 30 | 30 | 30 | 209 |

| | Average within-bullet correlation matrix | | | | |
|---|---|---|---|---|---|
| | Naa-As | ICP-Sb | ICP-Cu | ICP-Bi | ICP-Ag |
| NAA-As | 1.00 | 0.05 | 0.04 | 0.03 | 0.04 |
| ICP-Sb | 0.05 | 1.00 | 0.67 | 0.32 | 0.31 |
| ICP-Cu | 0.04 | 0.67 | 1.00 | 0.26 | 0.30 |
| ICP-Bi | 0.03 | 0.32 | 0.26 | 1.00 | 0.16 |
| ICP-Ag | 0.04 | 0.31 | 0.30 | 0.16 | 1.00 |

covariance matrices were estimated separately for each manufacturer, on both the raw (untransformed) and log-transformed scales, for Sb, Cu, Bi, and Ag (measured with ICP-OES by all four manufacturers) and As (measured with NAA by all four manufacturers). Only the variances and covariances as calculated on the log scale are shown in Table K.4 because the square roots of the variances (diagonal terms) are estimates of the RSD. (These RSDs differ slightly from those cited in Table 2.2 in Chapter 2.) The within-bullet covariance matrices are pooled (averaged) across manufacturer, and the correlation matrix is derived in the usual way: correlation between elements *i* and *j* equals the covariance divided by the product of the SDs; that is, $s_{jl} / \sqrt{s_{jj} s_{ll}}$ . (The correlation matrix based on the untransformed data is very similar.) As and Sn were also measured with ICP-OES on only the Federal bullets, so the 6 × 6 within-bullet variances and covariances, and the within-bullet correlations among the six measurements, are given in Appendix F.

The estimated correlation matrix indicates usually small correlations between the errors in measuring elements. Four notable exceptions are the correlation between the errors in measuring Sb and Cu, estimated as 0.67, and the correlations between the errors in measuring Ag and Sb, between Ag and Cu, and between Sb and Bi, all estimated as 0.30–0.32.

Figure K.1 demonstrates that association with plots of the three Cu measurements versus the three Sb measurements centered at their mean values, so (0, 0) is roughly in the center of each plot for 20 randomly selected bullets from one of the four boxes from CCI (Ref. 2). In all 20 plots, the three points increase from left to right. A plot of three points does not show very much, but one would not expect to see all 20 plots showing consistent directions if there were no association in the measurement errors of Sb and Cu. In fact, for all four manufacturers,

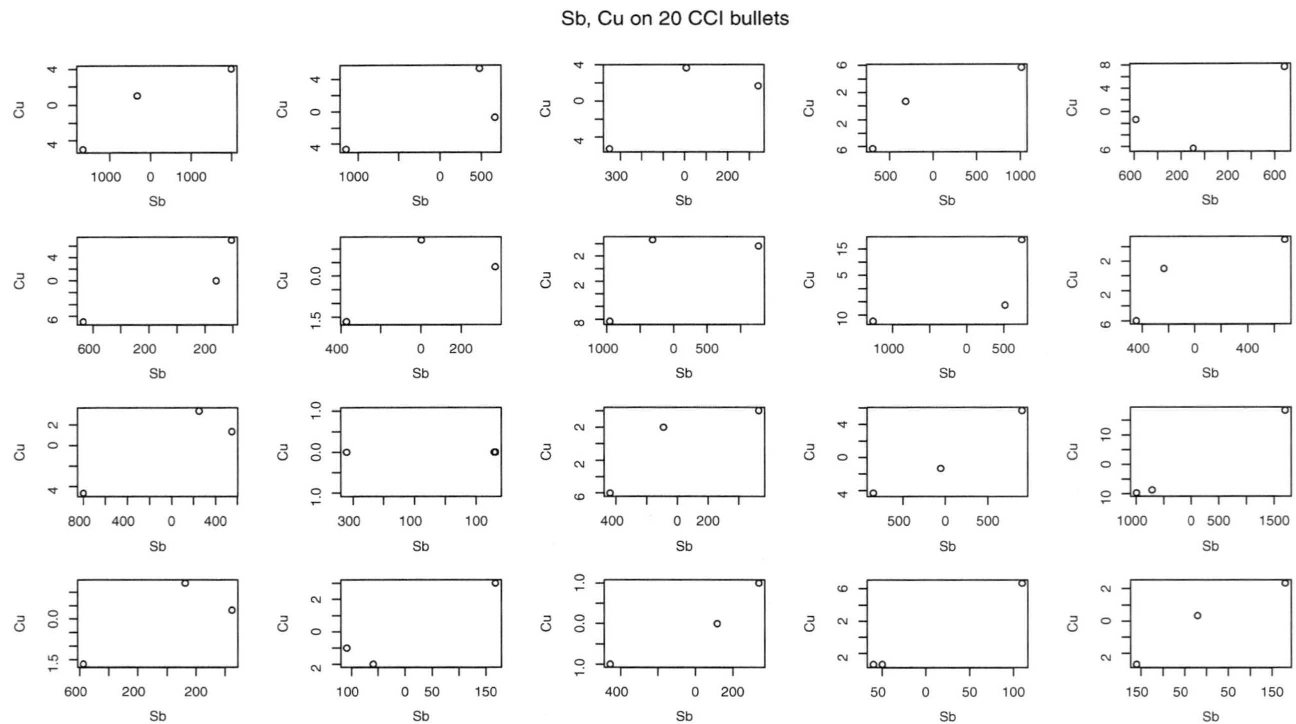


FIGURE K.1 Plots, for 20 CCI bullets, of three Cu measurements vs three Sb measurements. Each plot is centered at origin; that is, each plot shows $x_{i,Cu} - x_{Cu}$ vs $x_{i,Sb} - x_{Sb}$. If, as was commonly believed, errors in measuring Sb and Cu were independent, one would have expected to see increasing trends in about half these plots and decreasing trends in the other half. All these plots show increasing trends; 150 of the total of 200 plots showed increasing trends.

the estimated correlation between the three measurements in each bullet was positive for over 150 of the 200 bullets; this indicates further that the errors in measuring Sb and Cu may be dependent.

It has been assumed that the errors in measuring the different elements are independent, but these data suggest that the independence assumption may not hold. The nonindependence will affect the overall false positive probability of a match based on all seven intervals.

### 1,837-Bullet Data Set

Estimates of correlations among all seven elements measured with ICP-OES is not possible with the 1,837-bullet data set because the three replicates have been summarized with sample means and SDs. However, this data set does provide some information on within-bullet variances (not covariances) by providing the SD of the three measurements. Pooled estimates of the RSD, from the 800-bullet data set, and the median value of the reported SD divided by the reported average from bullets in the 1,837-bullet datas ets, are given in Table K.5. (Pooled RSDs are recommended for the alternative tests described in Section.4.) Because the three fragment averages (measurements) were virtually identical for several bullets, leading to sample SDs of 0, the FBI replaced these values as indicated in the notes that accompanied this data set (Ref. 6): "for those samples for which the three replicate concentration measurements for an element were so close to the same value that a better precision was indicated than could be expected from the ICP-OES procedure, the measured precision was increased to no less than the method precision." These values for the precision are also listed in Table K.5, in the third row labeled "Minimum SD (FBI)." The complete data set with 71,000+ bullets should be analyzed to verify the estimates of the uncertainty in the measurement errors and the correlations among them. (Note: All RSDs are based on ICP-OES measurements. RSDs for As and Sn are based on 200 Federal bullets. RSDs for Sb, Bi, Cu, and As are based on within-bullet variances averaged across four manufacturers (800 bullets); compare Table K.4. The estimated RSD for NAA-As is 5.1%.)

TABLE K.5 Pooled Estimates of Within-Bullet Relative Standard Deviations of Concentrations

| | As | Sb | Sn | Bi | Cu | Ag | Cd |
|---|---|---|---|---|---|---|---|
| 800 bullets, % | 4.3 | 2.1 | 3.3 | 4.3 | 2.2 | 4.6 | — |
| 1,837 bullets, 100 × med(SD/ave),% | 10.9 | 1.5 | 118.2 | 2.4 | 2.0 | 2.0 | 33.3 |
| Minimum SD (FBI) | 0.0002 | 0.0002 | 0.0002 | 0.0001 | 0.00005 | 0.00002 | 0.00001 |

## 2.4 Between-Bullet Variances and Covariances

The available data averages from the 1,837-bullet data set are plotted on a log scale in Figure K.2. To distinguish better the averages reported as "0.0000," log(0) is replaced with log(0.00001) = –11.5 for all elements except Cd, for which log(0) is replaced with log(0.000001) = –13.8. The data on Sb and Sn appear to be bimodal, and data on Cd before the 1,030$^{th}$ bullet (before the year 1997) are missing. The last panel (h) of the figure is a plot of the log(Ag) values only for log values between –7 (9e-4) and –5 (67e-4). This magnification shows a slight increase in Ag concentrations over time that is consistent with the findings noted by the FBI (Ref. 9).

Figure K.3 shows all pairwise plots of average concentrations in the 1837-bullet data set. Each plot shows the logarithm of the average concentration of an element versus the logarithm of the average concentration of each of the other six elements (once as an ordinate and once as an abscissa). Vertical and horizontal stripes correspond to missing or zero values that were replaced with values of log(1e-6) or log(1e-7). The plots of Sn vs Ag, As vs Sn, and Ag vs Bi show that some relationships between the bullet concentrations of these pairs of elements may exist. The data on Sn fall into two categories: those whose log (mean Sn concentration) is less than or greater than –5 (Sn less than or greater than 0.0067 ppm). The data on Sb fall into perhaps four identifiable subsets: those whose log (mean Sb concentration) is less than –1 (Sb concentrations around 0.0150 ppm, from 0.0001 to 0.3491 ppm), between –1 and 0 (Sb around 0.7 ppm, from 0.35 to 1 ppm), between 0 and 1 (Sb around 1.6 ppm, from 1.00 to 2.17 ppm), and greater than 1 (Sb around 3 ppm, from 2.72 to 10.76 ppm), perhaps corresponding to "soft," "medium," "hard," and "very hard" bullets.

If the 1,837-bullet data set were a random sample of the population of bullets, an estimate of the correlation (linear association) between two elements—say, Ag and Sb—is given by the Pearson sample correlation coefficient:

$$\frac{\sum_{k=1}^{1837}(\bar{x}_{Ag,k}-\bar{x}_{Ag,\cdot})(\bar{x}_{Sb,k}-\bar{x}_{Sb,\cdot})}{\left[\sum_{k=1}^{1837}(\bar{x}_{Ag,k}-\bar{x}_{Ag,\cdot})^2\sum_{k=1}^{1837}(\bar{x}_{Sb,k}-\bar{x}_{Sb,\cdot})^2\right]^{1/2}} \tag{2}$$

where again the $x$'s refer to the logarithms of the concentrations, for example, $\bar{x}_{Ag,k}$ is the logarithm of the mean concentration of Ag in bullet $k$, and $\bar{x}_{Ag,\cdot}$ is the average $\sum_{k=1}^{1837}\bar{x}_{Ag,k}/1{,}837$. For other pairs of elements, the number 1,837 is replaced with the number of bullets in which both elements are measured. (Robust estimates of the correlations can be obtained by trimming any terms in the summation that appear highly discrepant from the others.) A nonparametric estimate of the linear association, Spearman's rank correlation coefficient, can be computed by replacing actual measured values in the formula above with their ranks (for example, replacing the smallest Sb value with 1 and the largest with 1,837).

(Ref. 10). Table K.6 displays the Pearson sample correlation coefficient from the 1,837-bullet data set. The Spearman correlations on the ranks on the 1,837-bullet data set, the number of data pairs of which both elements were nonmissing, and the Spearman rank correlation coefficient on the 1,373-bullet subset (with no missing values) are given in Appendix F; the values of the Spearman rank correlation coefficients are very consistent with those shown in Table K.6. All three sets of correlation coefficients are comparable in magnitude for nearly all pairs of elements, and all are positive. However, because the 1,837-bullet data set is not a random sample, no measures of statistical significance are attributed to any correlation coefficients. The values are useful primarily for relative comparisons between correlation coefficients computed in this table.

### 2.5 Analysis of Randich et al. Data Set: Issues of Homogeneity

The data in Randich et al. (Ref. 7) were collected to assess the degree of inhomogeneity in lots of wires from which bullets are manufactured. Appendix H presents an analysis of those data. Here we only compare the within-replicate variances obtained on the 800-bullet data set with the within-lot variances in the Randich data. The former includes only five elements (As with NAA and Sb, Cu, Bi, and Ag with ICP), so variances on only these five elements are compared. As recommended earlier, these variances are calculated on the logarithms of the data, so they can be interpreted as the squares of the RSDs on the original scale.

For the As and Sb concentrations, the variability of the three measurements (beginning, middle, and end, or B, M, and E) is about the same as the variability of the three measurements in the bullets in the 800-bullet data set. For Bi and Ag, the within-lot variability (B, M, and E) is much smaller than the within-bullet variability in the 800-bullet data set. The within-lot variance of the three Cu measurements is considerably larger than the within-bullet variance obtained in the 800-bullet data set because of some very unusual measurements in five lots; when these lots are excluded, the estimated within-lot variance is comparable with the within-bullet variance in the 800-bullet data set. Randich et al. do not provide replicates or precise within-replicate measurement standard errors, so one cannot determine whether the precision of one of their measurements is equivalent to the precision of one of the FBI measurements. A visual display of the relative magnitude of the lot-to-lot variability (different lots) compared with the within-lot variability (B, M, and E) is shown in Figure K.4, which plots the log(measurement) by element as a function of lot number (in three cases, the lot number was modified slightly to avoid duplicate lot numbers, solely for plotting purposes: 424*A* $\rightarrow$ 425; 457 $\rightarrow$ 458; 456*A* $\rightarrow$ 457). Lot-to-lot variability is usually 9–12 times greater than within-lot variability: separate two-way analyses of variances on the logarithms of the measurements on the six elements, with the two factors "lot" (27 degrees of freedom for 28 lots) and "position in lot" (2

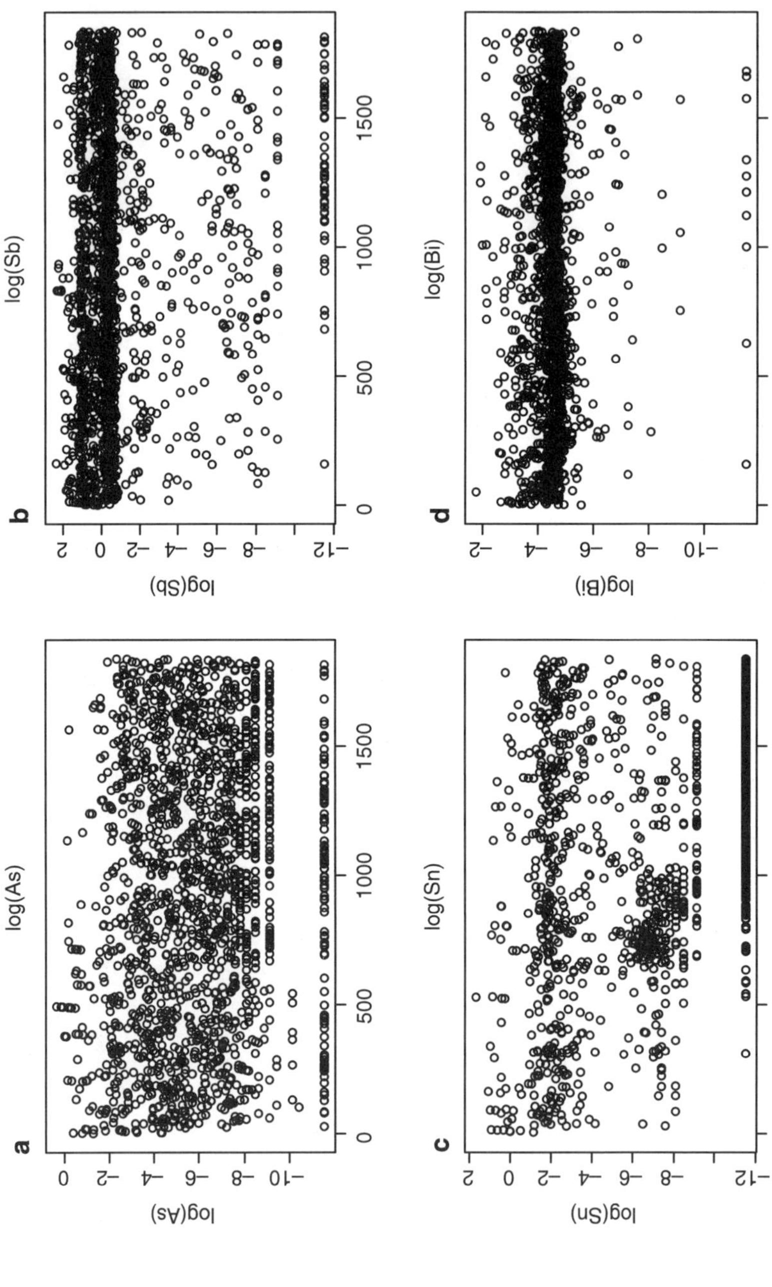
a
log(As)
b
log(Sb)
c
log(Sn)
d
log(Bi)

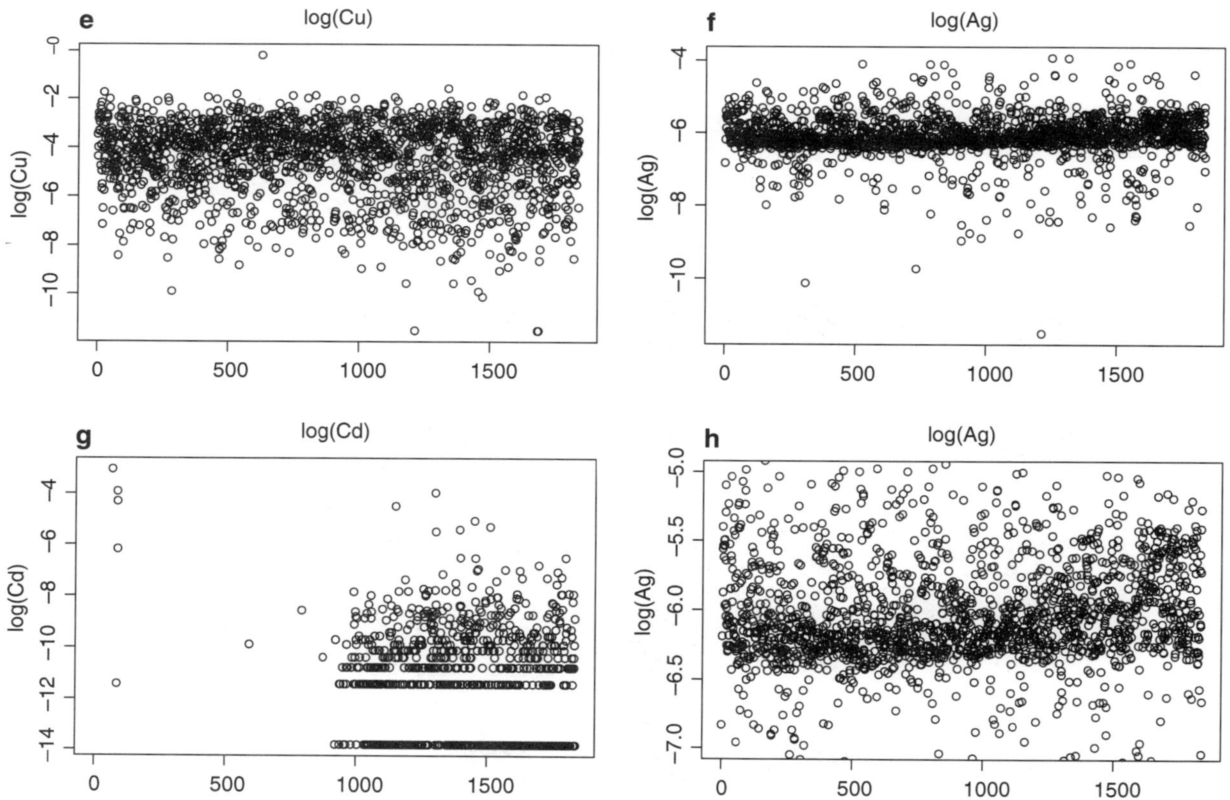


FIGURE K.2 Plots of log(mean concentrations), over time, in bullets from 1,837-bullet data set. (a) As; (b) Sb; (c) Sn; (d) Bi; (e) Cu; (f) Ag; (g) Cd; and (h) Ag, restricted to values between 0.0009 and 0.0067 (note slight increasing trend over time).

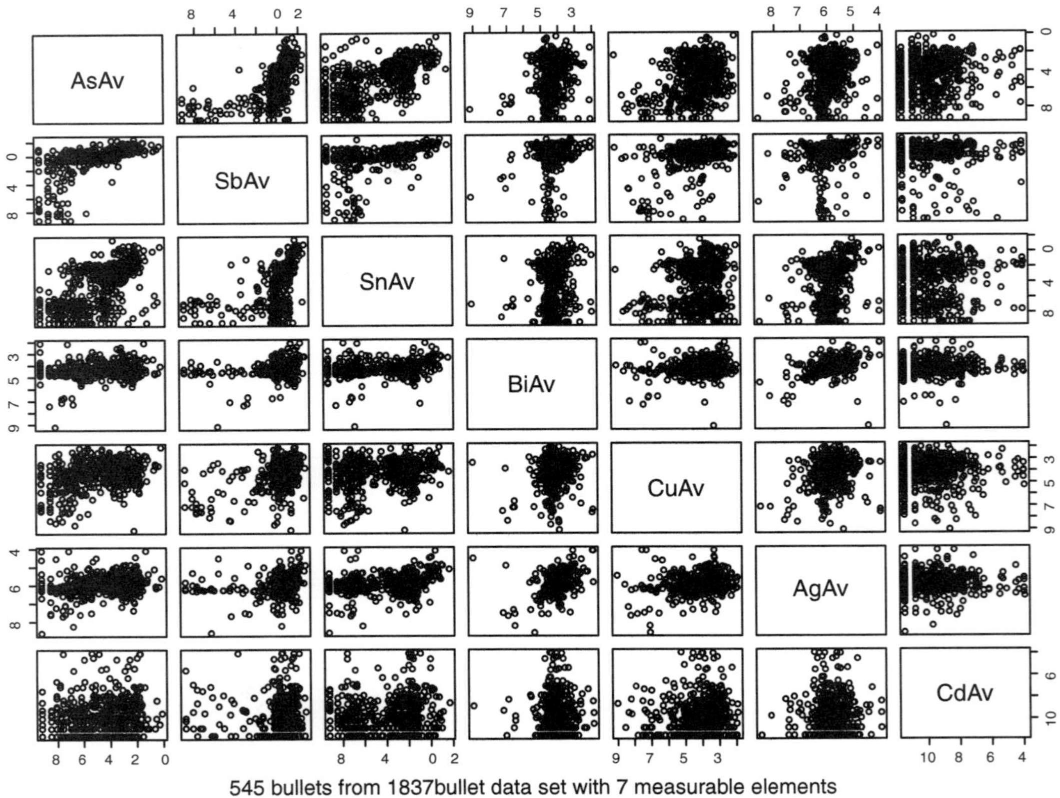


FIGURE K.3 Pairwise plots of log(mean concentrations) in bullets from the 1,837-bullet data set. Each pair of elements (such as Sb and As) is shown twice, once as Sb vs As and once as As vs Sb.

TABLE K.6 Between-Element Correlations[a] (1,837-Bullet Data Set)

| | As | Sb | Sn | Bi | Cu | Ag | Cd |
|---|---|---|---|---|---|---|---|
| As | 1.00 | 0.56 | 0.62 | 0.15 | 0.39 | 0.19 | 0.24 |
| Sb | 0.56 | 1.00 | 0.45 | 0.16 | 0.36 | 0.18 | 0.13 |
| Sn | 0.62 | 0.45 | 1.00 | 0.18 | 0.20 | 0.26 | 0.18 |
| Bi | 0.15 | 0.16 | 0.18 | 1.00 | 0.12 | 0.56 | 0.03 |
| Cu | 0.39 | 0.36 | 0.20 | 0.12 | 1.00 | 0.26 | 0.11 |
| Ag | 0.19 | 0.18 | 0.26 | 0.56 | 0.26 | 1.00 | 0.08 |
| Cd | 0.24 | 0.13 | 0.18 | 0.03 | 0.11 | 0.08 | 1.00 |

[a]Pearson correlation; see Equation 2. Spearman rank correlations are similar; see Appendix F.

TABLE K.7 Comparison of Within-Bullet and Within-Lot Variances[a]

| | ICP-As | ICP-Sb | ICP-Cu | ICP-Bi | ICP-Ag |
|---|---|---|---|---|---|
| Between lots: Randich et al. | 4,981.e-04 | 40.96e-04 | 17890e-04 | 60.62e-04 | 438.5e-04 |
| Within-bullet: 800-bullet data | 26.32e-04[b] | 4.28e-04 | 4.73e-04 | 18.25e-04 | 20.88e-04 |
| Within-lot: Randich et al. | 31.32e-04 | 3.28e-04 | 8.33e-04 | 0.72e-04 | 3.01e-04 |
| Ratio of within-lot to within-bullet: | 1.2 | 0.8 | 1.8 | 0.04 | 0.14 |

[a]Within-lot variance for Cu (line 3) is based on 23 of the 28 lots, excluding lots 423, 426, 454, 464, 465 (highly variable). The within-lot variance using all 28 lots is 0.0208.
[b]Based on NAA-As.

degrees of freedom for three positions: B, M, and E) confirm the nonsignificance of the position factor for all six elements—all except Sn—at the $\alpha$ level of significance. The significance for Sn results from two extreme values in this data set, both occurring at location E, on lot 424 (B = M = 414 and E = 21) and on lot 454 (B = 377, M = 367, and E = 45). Some lots also yielded three highly dispersed Cu measurements, for example, lot 465 (B = 81, M = 104, and E = 103) and lot 454 (B = 250, M = 263 and E = 156). In general, no consistent patterns (such as, $B < E < M$ or $E < M < B$) are discernible for measurements within lots on any of the elements, and, except for five lots with highly dispersed Cu measurements, the within-lot variability is about the same as or smaller than the measurement uncertainty (Appendix G).

### 2.6 Differences in Average Concentrations

The 1,837-bullet data set and the data in Table 1 of Randich et al. (Ref. 7)

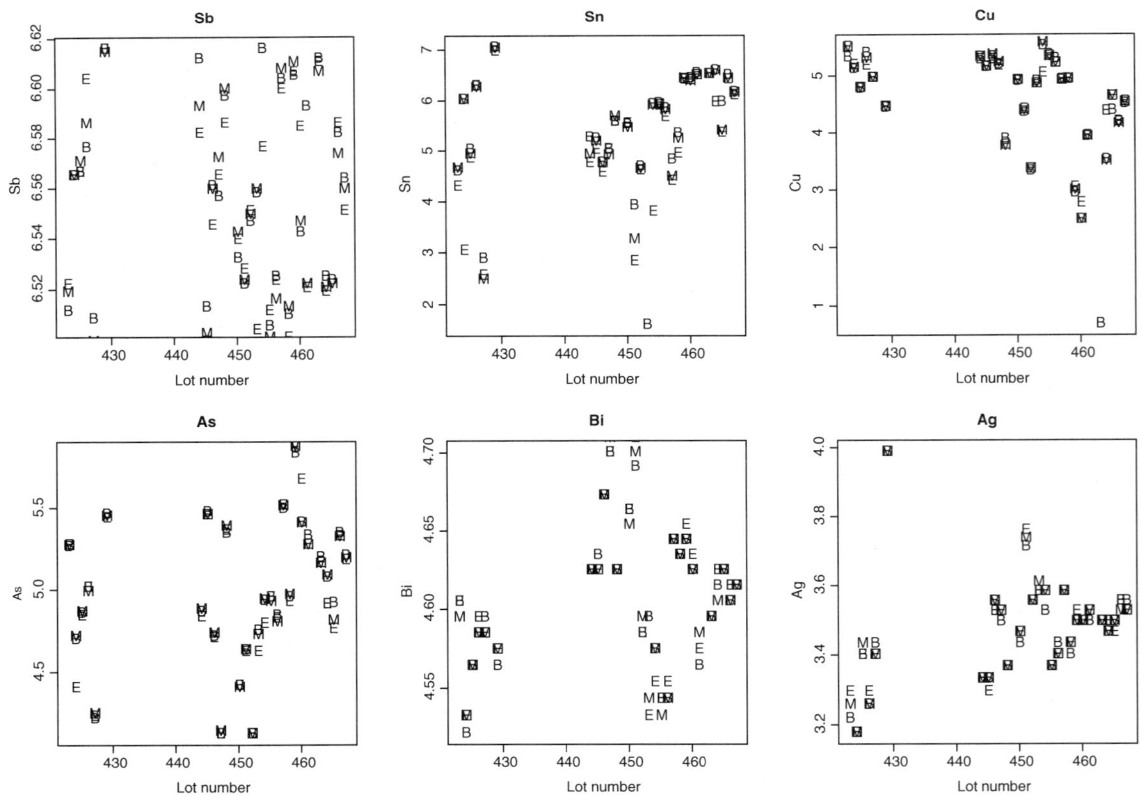


FIGURE K.4 Plot of log(element concentration) as function of lot number for data in Table 1 of Randich et al. (2002). In each panel, characters B, M, and E correspond to measurement taken at beginning, middle, and end of wire.

provide information on differences in average concentrations between bullets from different lots (in the case of Randich et al.) or sources (as suggested by the FBI for the 1,837-bullet data set). The difference in the average concentration relative to the measurement uncertainty is usually quite large for most pairs of bullets, but it is important to note the instances in which bullets come from different lots but the average concentrations are close. For example, lots 461 and 466 in Table 1 of Randich et al. (Ref. 7) showed average measured concentrations of five of the six elements within 3–6% of each other:

| | Sb | Sn | Cu | As | Bi | Ag |
|---|---|---|---|---|---|---|
| 461 (average) | 696.3 | 673.0 | 51.3 | 199.3 | 97.0 | 33.7 |
| 466 (average) | 721.0 | 632.0 | 65.7 | 207.0 | 100.3 | 34.7 |
| % difference | –3.4% | 6.4% | –21.8% | –3.7% | –3.3% | –2.9% |

Those data demonstrate that two lots may differ by as little as a few percent in as many as five (or even six, including Cd also) of the elements currently being measured in CABL analysis.

Further evidence of the small differences that can occur between the average concentrations in two apparently different bullets arises in 47 pairs of bullets, among the 854 bullets in the 1837-bullet data set in which all seven elements were measured (364,231 possible pairs). The 47 pairs of bullets matched by the FBI's 2-SD-overlap method are listed in Table K.8. For 320 of the 329 differences between elemental concentrations (47 bullet pairs × 7 elements = 329 element comparisons), the difference is within a factor of 3 of the measurement uncertainty. That is, if $\delta$ is the true difference in mean concentrations (estimated by the difference in the measured averages) and $\sigma$ = measurement uncertainty (estimated by a pooled SD of the measurements in the two bullets or root mean square of the two SDs), an estimate of $\delta/\sigma \leq 3$ is obtained on 320 of the 329 element differences. Table K.8 is ordered by the maximal (over seven elements) relative mean difference, or *RMD* (i.e., difference in sample means, divided by the larger of the two SDs). For the first three bullet pairs listed in Table K.8, $RMD \leq 1$ for all seven elements; for the next five bullet pairs, $RMD \leq 1.5$ for all seven elements; for 30 bullet pairs, the maximal *RMD* was between 2 and 3; and for the last nine pairs in the table, *RMD* was between 3 and 4. So, although the mean concentrations of elements in most of these 854 bullets differ by a factor that is many times greater than the measurement uncertainty, some pairs of bullets (selected by the FBI to be different) show mean differences that can be as small as 1 or 2 times the relative measurement uncertainty. This information on apparent distances between element concentrations relative to measurement uncertainty is used later in the recommendation for the equivalence $t$ test (see Section K.4).

TABLE K.8 Comparisons of 47 Pairs of Bullets from Among 854 of 1,837 Bullets Having Seven Measured Elements, Identified as Match by 2-SD-Overlap Method

| | | | | | (Difference in Mean Concentration)/SD | | | | | | | |
|---|---|---|---|---|---|---|---|---|---|---|---|---|
| | Bullet 1 | | Bullet 2 | | Elements | | | | | | | |
| | No. | Case | No. | Case | As | Sb | Sn | Bi | Cu | Ag | Cd | FPP[a] |
| 1 | 1,044 | 630 | 1,788 | 982 | 0.50 | 0.50 | 0.0 | 0.67 | 0.90 | 0.71 | 0.00 | 0.85 |
| 2 | 591 | 377 | 1,148 | 679 | 0.50 | 0.79 | 0.0 | 0.20 | 0.85 | 1.00 | 0.00 | 0.85 |
| 3 | 1,607 | 895 | 1,814 | 994 | 1.00 | 0.00 | 0.0 | 0.67 | 0.60 | 0.22 | 1.00 | 0.82 |
| 4 | 1,211 | 709 | 1,412 | 808 | 0.25 | 0.09 | 0.0 | 0.17 | 0.28 | 0.53 | 1.12 | 0.88 |
| 5 | 1,133 | 671 | 1,353 | 786 | 0.00 | 0.50 | 0.0 | 1.25 | 1.20 | 0.14 | 1.00 | 0.85 |
| 6 | 1,085 | 653 | 1,180 | 697 | 0.33 | 0.50 | 0.0 | 1.00 | 1.40 | 1.20 | 0.00 | 0.85 |
| 7 | 1,138 | 674 | 1,353 | 786 | 0.50 | 0.50 | 0.0 | 0.00 | 0.83 | 1.43 | 0.00 | 0.88 |
| 8 | 1,044 | 630 | 1,785 | 982 | 0.50 | 1.50 | 0.0 | 1.00 | 0.89 | 1.25 | 0.00 | 0.72 |
| 9 | 937 | 570 | 981 | 594 | 1.00 | 2.00 | 0.5 | 2.00 | 0.41 | 1.00 | 1.00 | 0.61 |
| 10 | 954 | 578 | 1,027 | 621 | 2.00 | 0.00 | 0.5 | 0.33 | 1.00 | 0.18 | 1.00 | 0.74 |
| 11 | 1,207 | 707 | 1,339 | 778 | 1.00 | 1.83 | 0.0 | 0.50 | 1.00 | 1.20 | 2.00 | 0.61 |
| 12 | 1,237 | 724 | 1,289 | 748 | 0.00 | 0.00 | 0.0 | 0.00 | 0.80 | 2.00 | 0.00 | 0.77 |
| 13 | 1,277 | 742 | 1,353 | 786 | 0.00 | 0.50 | 0.0 | 2.00 | 1.40 | 0.43 | 0.00 | 0.77 |
| 14 | 1,286 | 746 | 1,458 | 827 | 1.00 | 0.61 | 0.5 | 1.20 | 0.78 | 0.00 | 2.00 | 0.70 |
| 15 | 1,785 | 982 | 1,788 | 982 | 0.00 | 2.00 | 0.0 | 0.00 | 0.25 | 0.00 | 0.00 | 0.79 |
| 16 | 954 | 578 | 1,793 | 982 | 2.00 | 0.00 | 0.5 | 0.33 | 1.92 | 2.18 | 1.00 | 0.55 |
| 17 | 953 | 577 | 1,823 | 997 | 2.00 | 0.84 | 0.5 | 0.60 | 2.20 | 0.94 | 2.00 | 0.52 |
| 18 | 953 | 577 | 1,075 | 648 | 2.00 | 2.23 | 0.5 | 1.80 | 1.66 | 1.71 | 1.00 | 0.40 |
| 19 | 1,220 | 715 | 1,353 | 786 | 0.00 | 0.50 | 0.0 | 2.25 | 2.17 | 0.57 | 1.00 | 0.63 |
| 20 | 1,339 | 778 | 1,353 | 786 | 1.50 | 0.00 | 0.0 | 1.75 | 0.60 | 2.29 | 2.00 | 0.47 |
| 21 | 1,202 | 703 | 1,725 | 955 | 2.00 | 2.36 | 0.0 | 0.00 | 1.73 | 2.00 | 0.00 | 0.49 |
| 22 | 953 | 577 | 1,067 | 644 | 2.00 | 0.46 | 0.5 | 0.40 | 2.41 | 1.53 | 1.00 | 0.55 |
| 23 | 1,251 | 729 | 1,314 | 760 | 0.50 | 2.41 | 0.0 | 0.71 | 1.80 | 0.76 | 0.00 | 0.63 |
| 24 | 1,550 | 871 | 1,642 | 912 | 0.50 | 0.00 | 0.0 | 2.00 | 2.07 | 2.50 | 2.00 | 0.49 |
| 25 | 1,001 | 608 | 1,276 | 742 | 0.50 | 2.65 | 0.0 | 0.00 | 2.20 | 0.50 | 1.00 | 0.48 |
| 26 | 1,207 | 707 | 1,353 | 786 | 2.00 | 1.83 | 0.0 | 1.50 | 2.67 | 1.43 | 0.00 | 0.35 |
| 27 | 1,353 | 786 | 1,749 | 968 | 0.50 | 0.50 | 0.0 | 1.00 | 2.80 | 1.71 | 0.00 | 0.48 |
| 28 | 1,226 | 719 | 1,723 | 955 | 2.00 | 0.81 | 0.0 | 2.00 | 2.91 | 0.86 | 1.00 | 0.39 |
| 29 | 953 | 577 | 1,335 | 774 | 0.50 | 0.66 | 0.0 | 0.60 | 0.22 | 1.00 | 3.00 | 0.53 |
| 30 | 954 | 578 | 1,173 | 692 | 1.50 | 0.00 | 0.5 | 3.00 | 2.62 | 0.27 | 0.00 | 0.31 |
| 31 | 1,120 | 666 | 1,315 | 761 | 2.00 | 0.00 | 0.0 | 3.00 | 0.78 | 1.00 | 2.00 | 0.40 |
| 32 | 1,133 | 671 | 1,138 | 674 | 0.50 | 0.00 | 0.0 | 1.67 | 1.83 | 3.00 | 1.00 | 0.41 |
| 33 | 1,138 | 674 | 1,207 | 707 | 1.67 | 2.00 | 0.0 | 3.00 | 1.83 | 0.00 | 0.00 | 0.36 |
| 34 | 1,244 | 725 | 1,569 | 881 | 0.00 | 1.82 | 0.0 | 2.00 | 2.27 | 3.00 | 0.00 | 0.36 |
| 35 | 1,245 | 726 | 1,305 | 757 | 0.50 | 0.86 | 0.0 | 0.50 | 2.33 | 1.43 | 3.00 | 0.47 |
| 36 | 1,245 | 726 | 1,518 | 859 | 1.00 | 0.48 | 0.0 | 3.00 | 0.67 | 0.00 | 0.00 | 0.55 |
| 37 | 1,630 | 907 | 1,826 | 998 | 2.33 | 0.87 | 0.0 | 2.00 | 2.09 | 3.00 | 1.00 | 0.34 |
| 38 | 1,709 | 947 | 1,750 | 969 | 1.00 | 0.50 | 0.0 | 3.00 | 0.79 | 2.20 | 2.00 | 0.40 |
| 39 | 921 | 563 | 1,015 | 615 | 0.50 | 3.00 | 0.0 | 1.00 | 3.13 | 3.00 | 1.00 | 0.22 |
| 40 | 1,138 | 674 | 1,749 | 968 | 0.00 | 0.00 | 0.0 | 1.33 | 3.17 | 0.67 | 0.00 | 0.55 |
| 41 | 1,277 | 742 | 1,429 | 816 | 1.67 | 1.14 | 0.0 | 0.50 | 3.20 | 1.00 | 0.00 | 0.47 |
| 42 | 1,220 | 715 | 1,277 | 742 | 0.00 | 0.00 | 0.0 | 0.50 | 3.33 | 2.33 | 1.00 | 0.48 |

TABLE K.8 *continued*

| | Bullet 1 | | Bullet 2 | | (Difference in Mean Concentration)/SD Elements | | | | | | | |
|---|---|---|---|---|---|---|---|---|---|---|---|---|
| | No. | Case | No. | Case | As | Sb | Sn | Bi | Cu | Ag | Cd | FPP[a] |
| 43 | 1,305 | 757 | 1,518 | 859 | 1.50 | 0.39 | 0.0 | 2.50 | 3.00 | 3.33 | 3.00 | 0.17 |
| 44 | 1,133 | 671 | 1,207 | 707 | 2.00 | 2.00 | 0.0 | 0.33 | 3.67 | 1.80 | 1.00 | 0.21 |
| 45 | 1,133 | 671 | 1,749 | 968 | 0.50 | 0.00 | 0.0 | 3.00 | 1.60 | 3.67 | 1.00 | 0.18 |
| 46 | 1,169 | 689 | 1,725 | 955 | 0.00 | 0.40 | 0.0 | 1.00 | 0.13 | 3.75 | 1.00 | 0.33 |
| 47 | 1,689 | 934 | 1,721 | 953 | 0.33 | 2.18 | 4.0 | 3.00 | 0.68 | 0.80 | 0.00 | 0.17 |

NOTE: Columns 1–4 give the case number and year for the two bullets being compared; columns As through Cd give values of the relative mean difference (RMD); that is, $(\bar{x}_j - \bar{y}_j)/\max(s_{xj}, s_{yj})$. Values less than 1 indicate that the measured mean difference in concentration is less than or equal to the measurement uncertainty (≈ 2–4% in most cases). The bullet pairs are listed in order of maximal RMD (over the seven elements). The maximal RMD is less than or equal to the measurement uncertainty (MU) for all seven elements for three comparisons (lines 1–3); less than or equal to 1.5 (MU) for eight comparisons (lines 1–8); between 2 (MU) and 3 (MU) for 30 comparisons (lines 9–38), and between 3 (MU) and 4 (MU) for seven comparisons (lines 39–47). The last column is the product of the apparent FPP of the FBI 2-SD-overlap procedure, assuming independence among measurement errors, based on Table K.9 (see Section 3.3).
[a]FPP = false-positive probability.

## 3. ESTIMATING FALSE-POSITIVE PROBABILITY

In this section, the false-positive probability (FPP) of the 2-SD-overlap and range-overlap procedures is estimated. The following notation will be used:

$x_{ijk}$ = $i^{th}$ measurement ($i$=1,2,3) of $j^{th}$ element ($j$ = 1,...,7) on $k^{th}$ CS bullet
$y_{ijk}$ = $i^{th}$ measurement ($i$=1,2,3) of $j^{th}$ element ($j$ = 1,...,7) on $k^{th}$ PS bullet

where "measurement" denotes an average (over triplicates) on one of the three pieces of the bullet (or bullet fragment). When the measurements are transformed with logarithms, $x_{ijk}$ will denote the log of the measurement (more likely to be normally distributed; see Section 3.2.2). To simplify the notation, the subscript $k$ is dropped. The mean and SD of the three measurements of a CS or PS bullet can be expressed as follows:

$\bar{x}_j = \sum_{i=1}^{3} \bar{x}_{ij} / 3 = (x_{1j} + x_{2j} + x_{3j}) / 3$ = sample mean of three measurements, element $j$, CS bullet

$s_{xj} = \left[\sum_{i=1}^{3} (x_{ij} - \bar{x}_j)^2 / 2\right]^{\frac{1}{2}}$ = SD of three measurements of element $j$ on CS bullet

$\bar{y}_j = {}^{3}_{i=1} y_{ij} / 3 = (y_{1j} + y_{2j} + y_{3j}) / 3$ = sample mean of three measurements, element $j$, PS bullet

$s_{yj} = \left[\sum_{i=1}^{3} (y_{ij} - \bar{y}_j)^2 / 2\right]^{\frac{1}{2}}$ = SD of three measurements of element $j$ on PS bullet

$(\bar{x}_j - 2s_{xj}, \bar{x}_j + 2s_{xj})$ = 2-SD interval for CS bullet

$(\bar{y}_j - 2s_{yj}, \bar{y}_j + 2s_{yj})$ = 2-SD interval for PS bullet

$(min(x_{1j}, x_{2j}, x_{3j}), max(x_{1j}, x_{2j}, x_{3j}))$ = range interval for CS bullet

$(min(y_{1j}, y_{2j}, y_{3j}), max(y_{1j}, y_{2j}, y_{3j}))$ = range interval for PS bullet

The sample means $\bar{x}_j$ and $\bar{y}_j$ are estimates of the true mean concentrations of element $j$ in the lead source from which the CS and PS bullets were manufactured, which will be denoted by $\mu_{x_j}$ and $\mu_{y_j}$, respectively. (The difference between the two means will be denoted $\delta_j$.) Likewise, the SDs $s_{x_j}$ and $s_{y_j}$ are estimates of the measurement uncertainty, denoted by $\sigma_j$. We do not expect the sample means $\bar{x}_j$ and $\bar{y}_j$ to differ from the true mean concentrations $\mu_{x_j}$ and $\mu_{y_j}$ by much more than the measurement uncertainty $(2 \cdot \sigma_j / \sqrt{3} \approx 1.15\sigma_j)$, but it is certainly possible (probability, about 0.10) that one or both of the sample means will differ from the true mean concentrations by more than $1.15\sigma_j$. Similarly, the sample mean difference, $\bar{x}_j - \bar{y}_j$, is likely (probability, 1.05) to fall within $1.96\sqrt{\sigma_j^2/3 + \sigma_j^2/3} = 1.6\sigma_j$ of the true difference $\mu_{xj} - \mu_{yj}$, and $\bar{x}_j - \bar{y}_j$ can be expected easily to lie within $3.5448\sigma_j$ of the true difference (probability, 0.9996). (Those probabilities are approximately correct if the data are lognormally distributed and the measurement error is less than 5%.)

The 2-SD interval (or the range interval) for the CS bullet can overlap with, or match, the 2-SD interval (or the range interval) for the PS bullet in any one of four ways—slightly left, slightly right, completely surrounds, and completely within—and can fail to overlap in one of two ways—too far left and too far right.

Because our judicial system is based on the premise that convicting an innocent person is more serious than acquitting a guilty person, we focus on the probability that two bullets match by either the 2-SD-overlap or range-overlap procedure, given that the mean concentrations of the elements are really different. We first describe the FBI's method of estimating the probability, and then we use simulation to estimate the FPP.

### 3.1 FBI Calculation of False-Positive Probability

The FBI reported an apparent FPP that was based on the 1,837-bullet data set (Ref. 11). The authors repeated the method on which the FBI's estimate was based as follows.

The 2-SD-overlap procedure is described in the analytical protocol (Ref. 11). Each bullet was compared with every other bullet by using the 2-SD-overlap criterion on all seven elements, or [(1,837)(1,836)/2] = 1,686,366 comparisons. Among these 1,837 bullets, 1,393 matched no other bullets. Recall that all seven elements were measured in only 854 bullets. In only 522 bullets, six elements were measured (Cd was missing in 519; and Sn was missing in 3). In 372 bullets, five elements were measured, and in 86 bullets, four were measured. The results showed that 240 bullets "matched" one other bullet, 97 "matched" two bullets, 40 "matched" three bullets, and 12 "matched" four bullets. Another 55 bullets "matched" anywhere from 5 to 33 bullets. (Bullet 112, from case 69 in 1990, matched 33 bullets, in part because only three elements—Sb, Ag, and Bi—were measured and were therefore eligible for comparison with only three elements in the other bullets.) A total of 1,386 bullets were found to have "matched" another bullet [240(1 bullet) + 97(2 bullets) + 40(3 bullets) + 12(4 bullets) + ... = 1,386], or 693 (= 1386/2) unique pairs of bullets matched. The FBI summarized the results by claiming an apparent FPP of 693/1,686,366, or 1 in 2,433.4 ("about 1 in 2,500").

That estimated FPP is probably too small, inasmuch as this 1,837-bullet data set is not a random sample of any population and may well contain bullets that tend to be further apart than one would expect in a random sample of bullets.

### 3.2 Simulating False-Positive Probability

We simulate the probability that the 2-SD interval (or range interval) for one bullet's concentration of one element overlaps with the 2-SD interval (or range interval) for another bullet's concentration of that element. The simulation is described below.

The CS average, $\bar{x}$, is an estimate of the true mean concentration, $\mu_x$; similarly, the PS average, $\bar{y}$, is an estimate of its true mean concentration, $\mu_y$. We simulate three measurements, normally distributed with mean $\mu_x = 1$ and measurement uncertainty $\sigma$, to represent the measurements of the CS bullet, and three measurements, normally distributed with mean $\mu_y = \mu_x + \delta$ and measurement uncertainty $\sigma$ to represent the measurements of the PS bullet, and determine whether the respective 2-SD intervals and range intervals overlap. We repeat this process 100,000 times, for various values of $\delta$ (0.1, 0.2, ..., 7.0) and $\sigma$ (0.005, 0.010, 0.015, 0.020, 0.025, and 0.030, corresponding to measurement uncertainty 0.5%, 1.0%, 1.5%, 2.0%, 2.5%, and 3.0% relative to $\mu = 1$), and we count the proportion of the 100,000 trials in which the 2-SD intervals or range intervals overlap. In this simulation, the measurement error is normally distributed. (Because $\sigma$ is small, 1.5–3.0%, the results with lognormally distributed error are virtually the same.) Unless $\delta = 0$, the FPPs for the two procedures should be small. We denote the two FPPs by $FPP_{2SD}(\delta,\sigma)$ and $FPP_{RG}(\delta,\sigma)$, respectively. Appendix F shows that the FPP is a function of only the ratio $\delta/\sigma$; that is, $FPP_{2SD}(1,1) = FPP_{2SD}(2,2) = FPP_{2SD}(3,3)$, and so on, and likewise for $FPP_{RG}(\delta,\sigma)$.

The FPP for the 2-SD-overlap method can be written 1 – P{no overlap}, where "P{ . . . }" denotes the probability of the event in braces. No 2-SD overlap occurs when either $\bar{x} + 2s_x < \bar{y} - 2s_y$ or $\bar{y} + 2s_y < \bar{x} - 2s_x$; that is, when either $(\bar{y} - \bar{x}) > 2(s_x + s_y)$ or $(\bar{x} - \bar{y}) > 2(s_x + s_y)$ or equivalently, when $|\bar{x} - \bar{y}| > 2(s_x + s_y)$. Thus, 2-SD overlap occurs whenever the difference between the two means is less than twice the sum of the two SDs on the two samples. (The average value of $s_x$ or $s_y$, the sample SD of three normally distributed measurements with true standard deviation $\sigma$, is $0.8862\sigma$, so on the average two bullets match in the 2-SD-overlap procedure whenever the difference in their sample means is within about $3.5448\sigma$.)

Likewise, no range overlap occurs when either $\max\{x_1,x_2,x_3\} < \min\{y_1,y_2,y_3\}$ or $\max\{y_1,y_2,y_3\} < \min\{x_1,x_2,x_3\}$. The minimum and maximum of three measurements in a normal distribution with measurement uncertainty $\sigma$ can be expected to lie within $0.8463\sigma$ of the true mean, so, very roughly, range overlap occurs on the average when the difference in the sample means lies within 0.8463 + 0.8463 = $1.6926\sigma$ of each other.

With measurement uncertainty (MU) equal to $\sigma$, the two probabilities are simulated (for only one element, so subscript $j$ is dropped for clarity):

$$FPP_{2SD}(\delta,\sigma) = 1 - P\{\text{"no overlap"}\} = 1 - P\{|\bar{x} - \bar{y}| > 2(s_x + s_y) \mid \mu_y - \mu_x = \delta, \text{MU} = \sigma\}$$

$$FPP_{RG}(\delta,\sigma) = 1 - P\{\max(y_1,y_2,y_3) < \min(x_1, x_2, x_3) \text{ or } \max(x_1,x_2,x_3) < \min(y_1,y_2,y_3) \mid \mu_y - \mu_x = \delta, \text{MU} = \sigma\}$$

where P{A|S} denotes the probability that A occurs (for example, "$|\bar{x} - \bar{y}| > 2(s_x + s_y)$" under conditions given by S (for example, "true difference in means is $\delta$, and the measurement uncertainty is $\sigma$"). The steps in the simulation algorithm follow. Set a value of $\delta$ (0.0, 0.1, 0.2, ..., 7.0) percent to represent the true mean difference in concentrations and a value of $\sigma$ (0.5, 1.0, 1.5, 2.0, 2.5, 3.0) percent to represent the true measurement uncertainty.

1. Generate three values from a normal distribution with mean 1 and standard deviation $\sigma$ to represent $x_1$, $x_2$, $x_3$, the three measured concentrations of an element in a CS bullet. Generate three values from a normal distribution with mean 1 + $\delta$ and standard deviation $\sigma$ to represent $y_1,y_2,y_3$, the three measured concentrations of an element on a PS bullet.

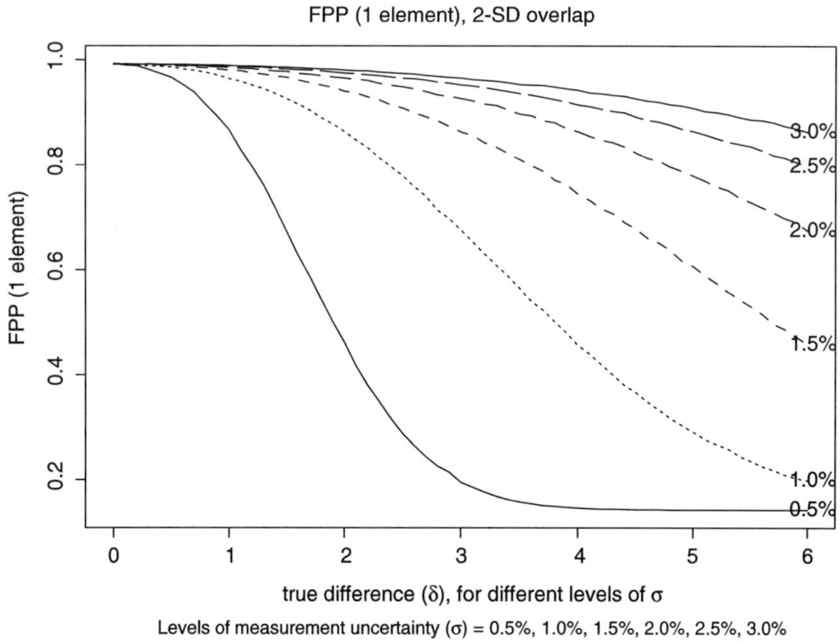


FIGURE K.5 Plot of estimated FPP for FBI 2-SD-overlap procedure as function of $\delta$ = true difference between (log)mean concentrations for single element. Each curve corresponds to different level of measurement uncertainty (MU) $\sigma$ (0.5%, 1.0%, 1.5%, 2.0%, 2.5%, and 3.0%).

2. Calculate $\bar{x}$, $\bar{y}$, $s_x$, and $s_y$, estimates of the means ($\mu_x$ and $\mu_y = 1 + \delta$) and SD ($\sigma$).

3. (a) For the 2-SD-overlap procedure:

$$\text{if } |\bar{x} - \bar{y}| > 2(s_x + s_y)\text{, record 0; otherwise record 1.}$$

(b) For the range-overlap procedure:

$$\text{if } \max\{x_1,x_2,x_3\} < \min\{y_1,y_2,y_3\} \text{ or } \max\{y_1,y_2,y_3\} < \min\{x_1,x_2,x_3\}\text{,}$$
$$\text{record 0; otherwise record 1.}$$

4. Repeat steps 1, 2, and 3 100,000 times. Estimate $FPP_{2SD}$ ($\delta$,$\sigma$) and $FPP_{RG}$ ($\delta$,$\sigma$) as the proportion of times that (a) and (b) record "1," respectively, in the 100,000 trials.

That algorithm was repeated for 71 values of $\delta$ (0.0, 0.001, ... , 0.070) and six values of $\sigma$ (0.005, 0.010, 0.015, 0.020, 0.025, and 0.030). The resulting estimates of the FPPs are shown in Figure K.5 ($FPP_{2SD}$) and Figure K.6 ($FPP_{RG}$)

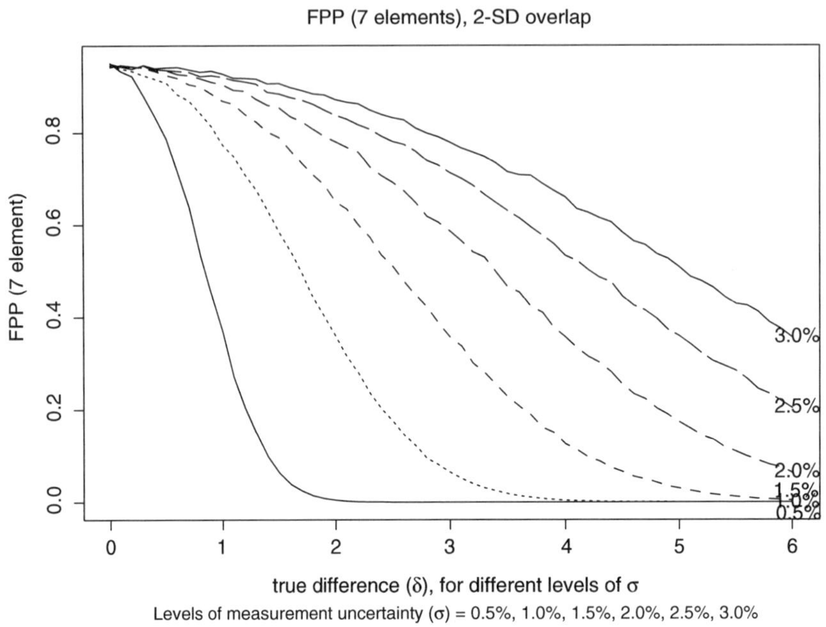


FIGURE K.6 Plot of estimated FPP for FBI 2-SD-overlap procedure as function of δ = true difference between (log)mean concentrations for seven elements, assuming independence among measurement errors. Each curve corresponds to different level of measurement uncertainty (MU) σ (0.5%, 1.0%, 1.5%, 2.0%, 2.5%, and 3.0%).

TABLE K.9 False-Positive Probabilities with 2-SD-Overlap Procedure (δ = 0–7%, σ = 0.5–3.0%)

| σ δ | 0 | 1 | 2 | 3 | 4 | 5 | 6 | 7 |
|---|---|---|---|---|---|---|---|---|
| 0.5 | 0.990 | 0.841 | 0.369 | 0.063 | 0.004 | 0.000 | 0.000 | 0.000 |
| 1.0 | 0.990 | 0.960 | 0.841 | 0.622 | 0.369 | 0.172 | 0.063 | 0.018 |
| 1.5 | 0.990 | 0.977 | 0.932 | 0.841 | 0.703 | 0.537 | 0.369 | 0.229 |
| 2.0 | 0.990 | 0.983 | 0.960 | 0.914 | 0.841 | 0.742 | 0.622 | 0.495 |
| 2.5 | 0.990 | 0.986 | 0.971 | 0.944 | 0.902 | 0.841 | 0.764 | 0.671 |
| 3.0 | 0.990 | 0.987 | 0.978 | 0.960 | 0.932 | 0.892 | 0.841 | 0.778 |

as a function of δ (true mean difference) for different values of σ (measurement uncertainty). Tables K.9 and K.10 provide the estimates for eight values of δ (0, 1, 2, 3, 4, 5, 6, and 7)% and six values of σ (0.5, 1.0, 1.5, 2.0, 2.5, and 3.0)%, corresponding roughly to observed measurement uncertainties of 0.5–3.0% (although some of the measurement uncertainties in both the 800-bullet data and the 1,837-bullet data were larger than 3.0%). The tables cover a wide range of values of δ/σ, ranging from 0 (true match) through 0.333 (δ = 1%, σ = 3%) to 14

TABLE K.10 False-Positive Probabilities with Range-Overlap Procedure δ = 0–7%, σ = 0.5–3.0%)

| σ δ | 0 | 1 | 2 | 3 | 4 | 5 | 6 | 7 |
|---|---|---|---|---|---|---|---|---|
| 0.5 | 0.900 | 0.377 | 0.018 | 0.000 | 0.000 | 0.000 | 0.000 | 0.000 |
| 1.0 | 0.900 | 0.735 | 0.377 | 0.110 | 0.018 | 0.002 | 0.000 | 0.000 |
| 1.5 | 0.900 | 0.825 | 0.626 | 0.377 | 0.178 | 0.064 | 0.018 | 0.004 |
| 2.0 | 0.900 | 0.857 | 0.735 | 0.562 | 0.377 | 0.220 | 0.110 | 0.048 |
| 2.5 | 0.900 | 0.872 | 0.792 | 0.672 | 0.524 | 0.377 | 0.246 | 0.148 |
| 3.0 | 0.900 | 0.882 | 0.825 | 0.735 | 0.626 | 0.499 | 0.377 | 0.265 |

(δ = 7%, σ = 0.5%). (Note: Only the value 0.900 for the range-overlap method when δ = 0 can be calculated explicitly without simulation. The simulation's agreement with this number is a check on the validity of the simulation.)

For seven elements, the 2-SD-overlap and range-overlap procedures declare a false match only if the 2-SD intervals overlapped on all seven elements. If the true difference in all element concentrations were equal (for example, δ = 2.0%), the measurement uncertainty was constant for all elements (for example, 2.0%), and the measurement errors for all seven elements were independent, the FPP for seven elements would equal the product of the per-element rate, seven times (for example, for δ = σ = 2%, $0.841^7 = 0.298$ for the 2-SD-overlap procedure, and $0.730^7 = 0.110$ for the range-overlap procedure). Figures K.7 and K.8, and Tables K.11 and K.12 give the corresponding FPPs, assuming independence among the measurement errors on all seven elements and assuming that the true mean difference in concentration is 100 δ percent.

The FPPs in Tables 3.11 and 3.12 are lower bounds because the analysis in the previous section indicated that the measurement errors may not be independent. (The estimated correlation between the errors in measuring Cu and Sb is 0.7, and the correlations between Sn and Sb, between Cu and Sn, between Ag and Cu, between Ag and Sb may be about 0.3.) The actual overall FPP is likely to be higher than $FPP^7$, probably closer to $FPP^6$ or $FPP^5$ [A brief simulation using the correlation matrix from the Federal bullets and assuming the Cd measurement is uncorrelated with the other 6 elements suggests that the FPP is closer to (per-element rate)$^5$]. To demonstrate that the FPP on seven elements is likely to be higher than the values shown in Table K.11 and K.12, we conducted another simulation, this time using actual data as follows:

1. Select one bullet from among the 854 bullets in which all seven elements were measured. Let $\mathbf{x}$ denote the vector of seven concentrations, and let $\mathbf{s}_x$ denote the vector of the seven SDs of the three measurements. (Note, only the mean and SD for a given bullet in this data set are given.)

2. Generate three values from a normal distribution with mean $\mathbf{x}$ and standard deviation $\mathbf{s}_x$ to represent $x_1, x_2, x_3$, the three measured concentrations of an

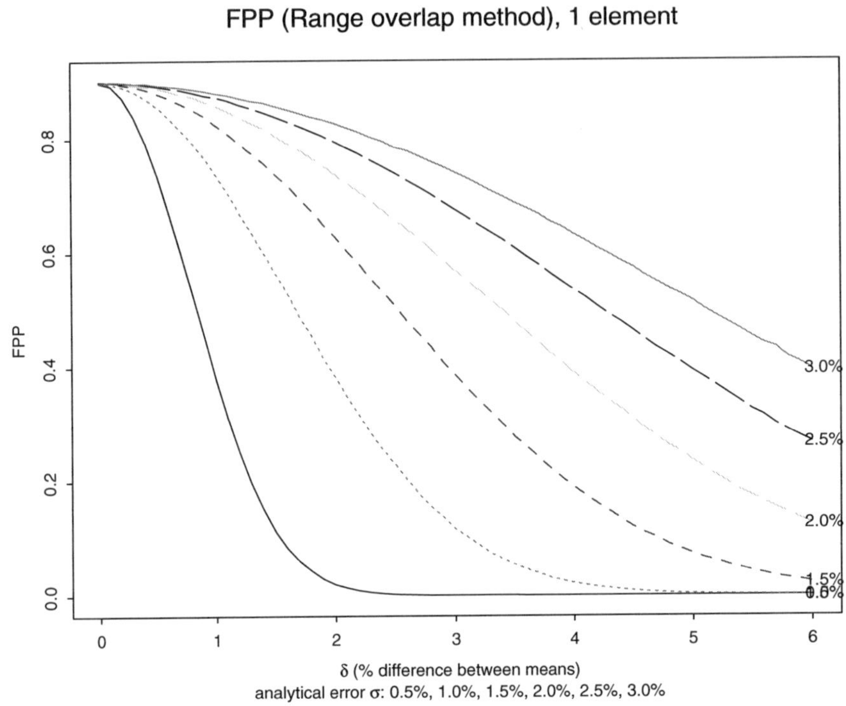


FIGURE K.7 Plot of estimated FPP for FBI range-overlap procedure as function of $\delta$ = true difference between (log)mean concentrations for single element. Each curve corresponds to different level of measurement uncertainty (MU) $\sigma$ (0.5%, 1.0%, 1.5%, 2.0%, 2.5% and, 3.0%).

element in the CS bullet. Generate three values from a normal distribution with mean $\mathbf{x}(1 + \delta)$ and SD $\mathbf{s}_x$ to represent $y_1$,$y_2$,$y_3$, the three measured concentrations of an element in the PS bullet. The three simulated $x$ values for element $j$ should have a mean close to the $j^{th}$ component of $\mathbf{x}$ ($j$ = 1, ..., 7) and SDs close to the $j^{th}$ component of $\mathbf{s}_x$. Similarly, the three simulated $y$ values for element $j$ should have a mean close to the $j^{th}$ component of $\mathbf{x}(1 + \delta)$ and SDs close to the $j^{th}$ component of $\mathbf{s}_x$.

3. Calculate $\bar{x}_j$, $\bar{y}_j$, $s_{xj}$, and $s_{yj}$, for $J$ = 1,…,7 elements, estimates of the means $\mathbf{x}$ and $(1 + \delta)\mathbf{x}$ and SD ($\mathbf{s}_x$).

4. For the 2-SD-overlap procedure:

if $|\bar{x}_j - \bar{y}_j| > 2(s_{xj} + s_{yj})$ for all seven elements, record 0; otherwise record 1.

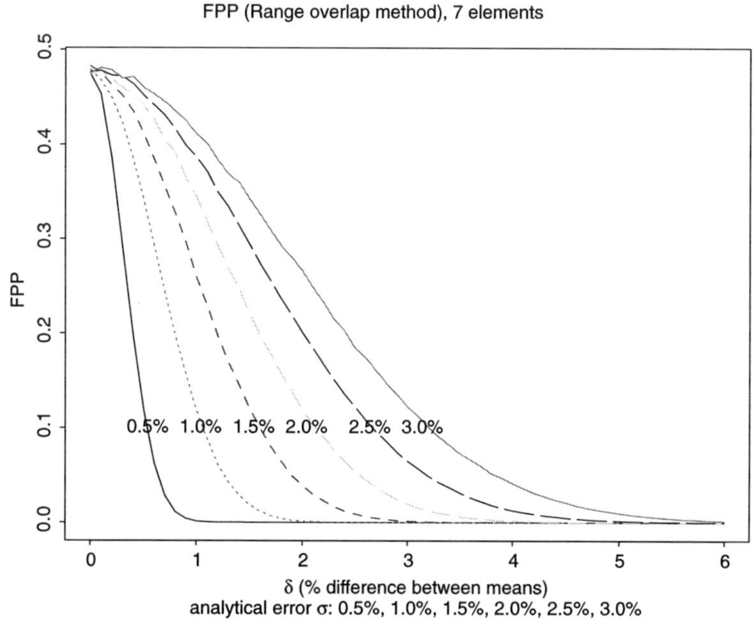


FIGURE K.8 Plot of estimated FPP for FBI range-overlap procedure as function of δ = true difference between (log)mean concentrations for seven elements, assuming independence among measurement errors. Each curve corresponds to different level of measurement uncertainty (MU) σ (0.5%, 1.0%, 1.5%, 2.0%, 2.5%, and 3.0%).

TABLE K.11 False-Positive Probabilities with 2-SD-Overlap Procedure, seven elements (assuming independence: δ = 0–7%, σ = 0.5–3.0%)

| σ δ | 0 | 1 | 2 | 3 | 4 | 5 | 6 | 7 |
|---|---|---|---|---|---|---|---|---|
| 0.5 | 0.931 | 0.298 | 0.001 | 0.000 | 0.000 | 0.000 | 0.000 | 0.000 |
| 1.0 | 0.931 | 0.749 | 0.298 | 0.036 | 0.001 | 0.000 | 0.000 | 0.000 |
| 1.5 | 0.931 | 0.849 | 0.612 | 0.303 | 0.084 | 0.013 | 0.001 | 0.000 |
| 2.0 | 0.931 | 0.883 | 0.747 | 0.535 | 0.302 | 0.125 | 0.036 | 0.007 |
| 2.5 | 0.931 | 0.903 | 0.817 | 0.669 | 0.487 | 0.302 | 0.151 | 0.062 |
| 3.0 | 0.931 | 0.911 | 0.850 | 0.748 | 0.615 | 0.450 | 0.298 | 0.175 |

TABLE K.12 False-Positive Probabilities with Range-Overlap Procedure, seven elements (assuming independence: δ = 0–7%, σ = 0.5–3.0%)

| σ δ | 0 | 1 | 2 | 3 | 4 | 5 | 6 | 7 |
|---|---|---|---|---|---|---|---|---|
| 0.5 | 0.478 | 0.001 | 0.000 | 0.000 | 0.000 | 0.000 | 0.000 | 0.000 |
| 1.0 | 0.478 | 0.116 | 0.001 | 0.000 | 0.000 | 0.000 | 0.000 | 0.000 |
| 1.5 | 0.478 | 0.258 | 0.037 | 0.001 | 0.000 | 0.000 | 0.000 | 0.000 |
| 2.0 | 0.478 | 0.340 | 0.116 | 0.018 | 0.001 | 0.000 | 0.000 | 0.000 |
| 2.5 | 0.478 | 0.383 | 0.197 | 0.062 | 0.011 | 0.001 | 0.000 | 0.000 |
| 3.0 | 0.478 | 0.415 | 0.261 | 0.116 | 0.037 | 0.008 | 0.001 | 0.000 |

For the range-overlap procedure:

if $\max\{x_{1j},x_{2j},x_{3j}\} < \min\{y_{1j},y_{2j},y_{3j}\}$ or $\max\{y_{1j},y_{2j},y_{3j}\} < \min\{x_{1j},x_{2j},x_{3j}\}$,

for all seven elements, record 0; otherwise record 1.

5. Repeat steps 1, 2, and 3 100,000 times. Estimate $FPP_{2SD}(\delta)$ and $FPP_{RG}(\delta)$ as the proportion of 1's that occur in step 4 in the 100,000 trials.

Four values of δ were used for this simulation—0.03, 0.05, 0.07, and 0.10, corresponding to 3%, 5%, 7%, and 10% differences in the means. If the typical relative measurement uncertainty is 2.0–3.0%, the results for 3%, 5%, and 7% should correspond roughly to the values in Tables K.11 and K.12 (2-SD-overlap and range-overlap, respectively, for seven elements), under columns headed 3, 5, and 7. The results of the simulations were:

| | δ | | | |
|---|---|---|---|---|
| method | 3.0% | 5.0% | 7% | 10% |
| with 2-SD overlap | 0.404 | 0.273 | 0.190 | 0.127 |
| with range overlap | 0.158 | 0.108 | 0.053 | 0.032 |

The FPP for the 2-SD-overlap method for all seven elements and δ = 3% is estimated in this simulation as 0.404, which falls between the two values in Table K.11 for σ = 1.5% (FPP, 0.303) and for σ 2.0% (FPP, 0.535). The FPP for the 2-SD-overlap method for all seven elements and δ = 5% is estimated in this simulation as 0.273, which falls between the two values in Table K.11 for σ = 2.0% (FPP, 0.125) and for σ = 2.5% (FPP, 0.302). The FPP for the 2-SD-overlap method for all seven elements and δ = 7% is estimated in this simulation as 0.190, which falls between the two values in Table K.11 for σ = 2.5% (FPP, 0.148) and for σ = 3.0% (FPP, 0.265). This simulation's FPPs for the range-overlap method for δ = 3%, 5%, and 7% result in estimates of the FPP as 0.158, 0.108, and 0.032, all of which correspond to values of σ greater than 3.0% in Table K.12 (columns for δ = 3, 5, and 7). The simulation suggests that measurement uncertainty may exceed 2–2.5%, and/or the measurement errors may be correlated.

Note that the FPP computation would be different if the mean concentrations differed by various amounts. For example, if the mean difference in three of the concentrations was only 1% and the mean difference in four of the concentrations was 3%, the overall FPP would involve products of the FPP(δ = 1%) and FPP(δ = 3%). The overall FPP is shown in Table K.8 on the basis of the observed mean difference/MU. Because most of the values of the RMD in Table K.8 are less than 3, the FPP estimates in the final column are high. The FPP estimates are effectively zero if the *RMD* exceeds 20% on two or more elements.

A separate confirmation of the FPPs in Table K.9 can be seen by using the apparent matches found between 47 pairs of bullets in Table K.8. Among all possible pairs of the 854 bullets from the 1,837-bullet data set (in which all seven elements were measured), 91 pairs showed a maximal RMD (difference in averages divided by 1 SD) across all seven elements of 4.0. The 2-SD-overlap procedure did not declare a match on these other 44 bullet pairs of the 91 pairs for which the maximal difference was 4%. Thus, the FPP could be estimated here as roughly 47/91, or 0.516. Table K.9 shows, for $\delta = 4\%$ and $\delta = 2.5\%$, an estimated FPP of 0.487. That is very close to the observed 0.516, although somewhat lower, possibly because of the correlation (lack of independence) that was used for the calculation from Table K.8 ($0.902^7 = 0.486$, but $0.902^{6.4} = 0.517$). Because homogeneous batches of lead, manufactured at different times, could by chance have the same chemical concentrations (within measurement error), the actual FPP could be even higher.

### 3.3 Chaining

The third method for assessing a match between bullets described in the FBI protocol [page 11, part (b)] has been called chaining. It involves the formation of "compositionally similar groups of bullets." We illustrate the effect of chaining on one bullet from the 1,837-bullet data set. According to the notes that accompanied this data set, "it might be most appropriate to consider all samples as unrelated or independent" (Ref. 10); thus, one would not expect to see compositional groups containing large numbers of bullets.

To see the effect of chaining, the algorithm (Ref. 1, p.11, part b; quoted in Section 3.1) was programmed. Consider bullet 1,044, from case 530 in 1997 in the 1,837-bullet data set. (Bullet 1044 is selected for no reason; any bullet will show the effect described below.) The measured elemental concentrations in that bullet are given in Table K.13. (According to Ref. 6, SDs for elements whose average concentrations were zero were inflated to the FBI's estimate of analytical uncertainty, noted in Table K.5 as "minimum SD (FBI).")

This bullet matched 12 other bullets; that is, the 2-SD interval overlapped on all elements with the 2-SD interval for 12 other bullets. In addition, each of the 12 other bullets matched other bullets; in total, 42 unique bullets were identified. The intervals for bullet 1,044 and the other 41 bullets are shown in Figure K.9a. The variability in the averages and the SDs of the 42 bullets would call into question the reasonableness of placing them all in the same compositional group. Bullets 150, 341, 634, and 647 clearly show much wider intervals than the others; even when eliminated from the set (Figure K.9b), a substantial amount of variability among the remaining bullets exists. The overall average and SD of the 42 average concentrations of the 42 "matching" bullets are given in the third and fourth lines of Table K.13 as "avg(42 avgs)" and "SD(42 avgs)." In all cases, the SDs are at least as large as, and usually 3–5 times larger than, the SD of bullet 1,044.

TABLE K.13 Statistics on bullet 1,044, to illustrate "Chaining" (see Section 3.4 and Figure K.9)

| | As | Sb | Sn | Bi | Cu | Ag | Cd |
|---|---|---|---|---|---|---|---|
| Avg | 0.0000 | 0.0000 | 0.0000 | 0.0121 | 0.00199 | 0.00207 | 0.00000 |
| SD | 0.0002 | 0.0002 | 0.0002 | 0.0002 | 0.00131 | 0.00003 | 0.00001 |
| Avg(42 Avgs) | 0.0004 | 0.0004 | 0.0005 | 0.0110 | 0.00215 | 0.00208 | 0.00001 |
| SD(42 Avgs) | 0.0006 | 0.0005 | 0.0009 | 0.0014 | 0.00411 | 0.00017 | 0.00001 |

Larger SDs lead to wider intervals and hence more matches. Using avg(42 avgs) ± 2SD(42 avgs) as the new 2-SD interval with which to compare the 2-SD interval from each of the 1,837 bullets results in a total of 58 matching bullets. (Even without the four bullets that have suspiciously wide 2-SD intervals, the algorithm yielded 57 matching bullets.) Although this illustration does not present a rigorous analysis of the FPP for chaining, it demonstrates that this method of assessing matches is likely to create even more false matches than either the 2-SD-overlap or the range-overlap procedure.

One of the questions presented to the committee (see Chapter 1) was, "Can known variations in compositions introduced in manufacturing processes be used to model specimen groupings and provide improved comparison criteria?" The authors of Ref. 8 (Carriquiry et al.) found considerable variability among the compositions in the 800-bullet data set; the analyses conducted here on the 1,837-bullet data set demonstrate that the variability in elemental compositions may be even greater than that seen in smaller data sets. Over 71,000 bullets have been chemically analyzed by the FBI during the last 15 years; thousands more will be analyzed, and millions more produced that will not be analyzed. In addition, thousands of statistical clustering algorithms have been proposed to identify groups in data with largely unknown success. For reasons outlined above, chaining, as one such algorithm, is unlikely to serve the desired purposes of identifying matching bullets with any degree of confidence or reliability. Because of the huge number of clustering algorithms designed for different purposes, this question on model specimen groupings posed to the committee cannot be answered at this time.

## 4. EQUIVALENCE TESTS

### 4.1 Concept of Equivalence Tests

Intuitively, the reason that the FPP could be higher than that claimed by the FBI is that the allowable range of the difference between the two sets of element concentrations is too wide. The FBI 2-SD-overlap procedure declares a match on an element if the mean difference in concentrations lies within twice the sum of the standard deviations; that is, if $|\bar{x}_j - \bar{y}_j| < 2(s_{xj} + s_{yj})$ for all $j = 1,2,\ldots, 7$

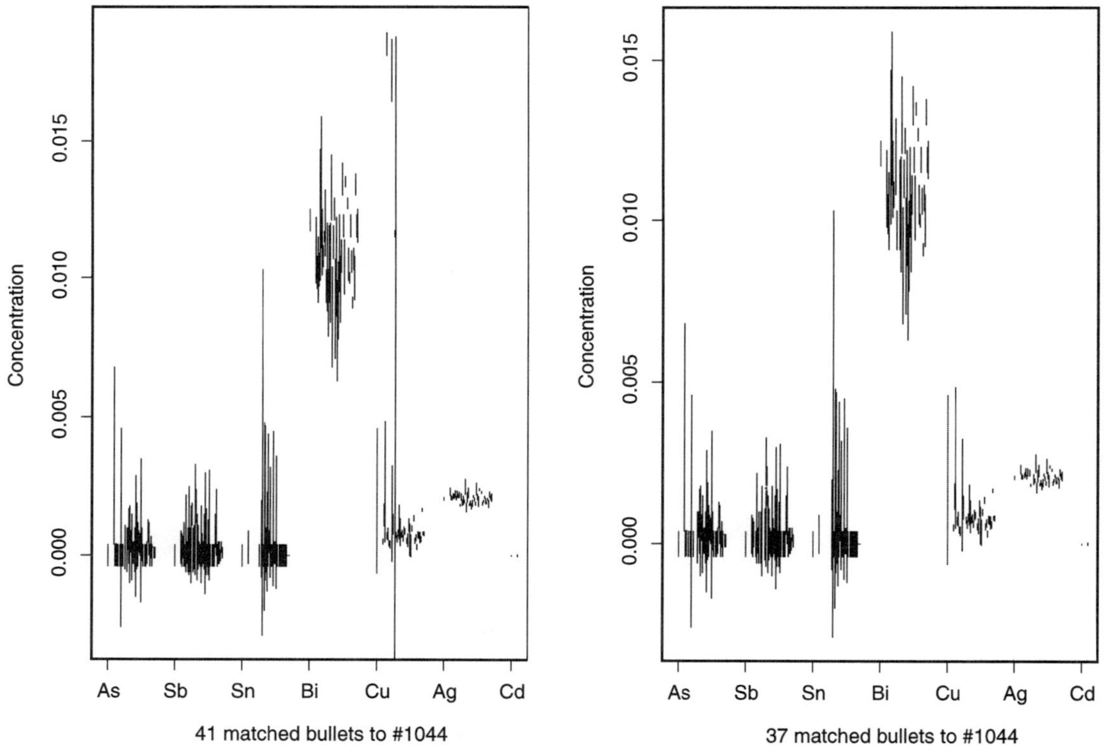


FIGURE K.9 Illustration of chaining. Panel (a) shows 2-SD-interval for bullet 1,044 (selected at random) as first line in each set of elements, followed by the 2-SD interval for each of 41 bullets whose 2-SD intervals overlap with that of bullet 1,044. Four of these 41 bullets had extremely wide intervals for Cu, so they are eliminated in Panel (b). Another 2-SD interval was constructed from SD of 42 (38) bullet averages on each element, resulting in a total of 58 (57) bullets that matched.

elements. The allowance used in the 2-SD interval, $2(s_{xj} + s_{yj})$ calculated for each element, is too wide for three reasons:

1. The measurement uncertainty in the difference between two *sample means*, each based on three observations, is $\sqrt{\sigma^2/3+\sigma^2/3} = 0.8165\sigma$. The average value of $s_{x_j} + s_{y_j}$, even when the measurements are *known* to be normally distributed, is $(0.8862\sigma + 0.8862\sigma) = 1.7724\sigma$, or roughly 2.17 times as large.

2. A sample SD based on only three observations has a rather high probability (0.21) of overestimating $\sigma$ by 25%, whereas a pooled SD based on 50 bullets each measured three times (compare Equation 2 in Appendix E) has a very small probability (0.00028) of overestimating $\sigma$ by 25%. (That is one of the reasons that the authors urge the FBI to use pooled SDs in its statistical testing procedures.)

3. The 2 in $2(s_{xj} + s_{yj})$ is about 2–2.5 times too large, assuming that

- The measurement uncertainty $\sigma$ is estimated by using a pooled SD.
- The procedure is designed to claim a match only if the true mean element concentrations differ by roughly the measurement uncertainty ($\delta \approx \sigma \approx 2 - 4\%$) or, at most, $\delta \approx 1.5\sigma \approx$ 3–6%. Measured differences in mean concentrations smaller than that amount would be considered analytically indistinguishable. Measured differences in mean concentrations larger than $\delta$ would be consistent with the hypothesis that the bullets came from different sources.

For these three reasons, the 2-SD interval claims a "match" for bullets that lie within an interval that is, on the average, about $3.5\sigma$ ($\sigma$ = measurement uncertainty), or about 7–17 percent. Hence, bullets whose mean concentrations differ by less than $3.5\sigma$ (about 7–17 percent) on all seven elements, have a high probability of being called "analytically indistinguishable."

The expected range of three normally distributed observations is $1.6926\sigma$, so the range-overlap method tends to result in intervals that are on average, about half as wide as the intervals used in the 2-SD-overlap procedure. This fact explains the results showing that the range-overlap method had a lower rate of false matches than the 2-SD-overlap method.

### 4.2 Individual Equivalence *t* Tests

An alternative approach is to set a per-element FPP of, say, 0.30 on any *one* element, so that the FPP on all seven elements is small, say, $0.30^5 = 0.00243$, or 1 in 412, to $0.30^6 = 0.000729$, or 1 in 1,372. This approach leads to an equivalence *t* test, which proceeds as follows:

1. Estimate the measurement uncertainty in measuring each element using a pooled SD, that is, the root mean square of the sample SDs from 50 to 100

bullets, where the sample SD on each bullet is based on the logarithms of the three measurements of each bullet. (The sample SDs on bullets should be monitored with a process-monitoring chart, called an *s-chart*; see Ref. 12, pages 76–78.) Denote the pooled SD for element $j$ as $s_{j,pool}$.

2. Calculate the mean of the logarithms of the three measurements of each bullet. Denote the sample means on element $j$ ($j = 1, 2, ..., 7$) for the CS and PS bullets as $\bar{x}_j$ and $\bar{y}_j$, respectively.

3. Calculate the difference between the sample means on each element, $\bar{x}_j - \bar{y}_j$. If they differ by less than 0.63 times $s_{j,pool}$ (about two-thirds of the pooled standard deviation for that element), for all seven elements, then the bullets are deemed "analytically indistinguishable (match)." If the sample means differ by less than 1.07 times $s_{j,pool}$ (slightly more than one pooled standard deviation for that element), for all seven elements, then the bullets are deemed "analytically indistinguishable (weak match)."

The limit 0.63 [or 1.07] allows for the fact that each sample mean concentration will vary slightly about its true mean (with measurement uncertainty $\sigma / \sqrt{3}$) and follows from the specification that (a) a false match on a single element has a probability of 0.30 and (b) a decision of "no match" suggests that the mean element concentrations are likely to differ by at least $1\sigma$ [or $1.5\sigma$], the uncertainty of a single measurement. That is, assuming that the uncertainty measuring a single element is 2.5 percent and the true mean difference between two bullet concentrations on this element is at least 2.5 percent [3.8 percent], then, with a probability of 0.30, caused by the uncertainty in the measurement process and hence in the sample means $\bar{x}_j$ and $\bar{y}_j$, the two sample means will, by chance, lie within $0.63s_{j,pool}$ [or 1.07] of each other, and the bullets will be judged as analytically indistinguishable on this one element (even though the mean concentrations of this element differ by 2.5%). A match occurs only if the bullets are analytically indistinguishable on all seven elements. Obviously, these limits can be changed, simply by choosing a different value for the per element false match probability, and a different value of $\delta$ (here $\delta = 1$ for a "match" and $\delta = 1.5$ for a "weak match.")

If the measurement errors in all elements were independent, then this procedure could be expected to have an overall FPP of $0.30^7 = 0.00022$, or about 1 in 4,572. The estimated correlation matrix in Section 3.3 suggests that the measurement errors are *not* all independent. A brief simulation comparing probabilities on 7 independent normal variates and 7 correlated normal variates (using the correlation matrix based on the Federal bullets given in Appendix F), indicated that the FPP is closer to $0.30^{5.2} = 0.002$, or about 1 in 500. To achieve the FBI's stated FPP of 0.0004 (1 in 2,500), one could use a per-element error rate of 0.222 instead of 0.30, because $0.222^{5.2} = 0.0004$. The limits for "match" and "weak match" would then change, from $0.636s_{j,pool}$ and $1.07s_{j,pool}$ to $0.47s_{j,pool}$ (about one-half of $s_{j,pool}$) and $0.88s_{j,pool}$, respectively. Table K.14 shows the calculations involved for the equivalence *t* tests on Federal bullets F001 and F002, using the data in Section 3.1 (log concentrations). The calculations are based on the pooled standard deviations using 200 Federal bullets (400 degrees of freedom; see Appendix F). Not all of the relative mean differences on elements (RMD = $(\bar{x}_j - \bar{y}_j)/s_{j,pool}$) are less than 0.86 in magnitude, but they are all less than 1.05 in magnitude. Hence the bullets would be deemed "analytically indistinguishable (weak match)."

The allowance $0.86s_{j,pool}$ can be written as $0.645s_{j,pool}\sqrt{2/3}$, and the value 0.645 arises from a noncentral *t* distribution (see Appendix F), used in an *equivalence t test* (Ref. 13), assuming that $n = 3$, that at least 100 bullets are used in the estimate $s_{j,pool}$ (200 bullets, or 400 degrees of freedom), and that mean concentrations with $\delta = \sigma$ (that is, within the measurement uncertainty) are considered analytically indistinguishable. The constant changes to $1.316\,s_{j,pool}\sqrt{2/3}$ = $1.07s_{j,pool}$ if one allows mean concentrations $\delta = 1.5\sigma$ to be considered "analytically indistinguishable." Other values for the constant are given in Appendix F; they depend slightly on *n* (here, three measurements per sample mean), on the number of bullets used to estimate the pooled variance (here, assumed to be at least 100), and, most importantly, upon the per-element-FPP (here, 0.30) and on $\delta/\sigma$ (here, 1–1.5). The choice of $\delta \approx \sigma$ used in the procedure is based on the observation that differences between mean concentrations among the seven elements ($\delta_j$, $j = 1,\ldots,7$) in three pairs of bullets in the 854-bullet subset of the 1,837-bullet data set (in which all seven elements were measured), which were assumed to be unrelated, can be as small as the measurement uncertainty ($\delta_j/\sigma_j \leq 1$ on all seven elements; compare Table K.8). Allowing matches between mean differences within 1.5, 2.0, or 3.0 times the measurement uncertainty increases the constant from 0.767 to 1.316, 1.925, or 3.147, respectively, and results in an increased allowance of the interval from $0.63s_{j,pool}$ ("match") to $1.07s_{j,pool}$ ("weak match"), $1.57s_{j,pool}$, and $2.57s_{j,pool}$, respectively (resulting in progressively weaker matches). The FBI allowance of $2(s_x + s_y) \approx 3.5448\sigma \approx 4.3415\,s_{j,pool}\sqrt{2/3}$, for the same per-element-FPP of 0.30 corresponds to $\delta/\sigma = 4.0$. That is, concentrations within roughly 4.3 times the measurement uncertainty would yield an FPP of roughly 0.30 on each element. (Because the measurement uncertainty on all 7 elements is roughly 2–5%, this corresponds to claiming that bullets are analytically indistinguishable whenever the concentrations lie within 8–20% of each other.) Those wide intervals resulted in 693 false matches among all possible pairs of the 1,837 bullets in the 1,837-bullet data set or in 47 false matches among all possible pairs of the 854 bullets in which all seven elements were measured. In contrast, using the limit $1.07s_{j,pool}$ resulted in zero matches among the 854 bullets.

The use of equivalence *t* tests for comparing two bullets depends only on a model for measurement error (lognormal distribution, or, if $\sigma/\mu$ is small, normal

TABLE K.14 Equivalence $t$-Tests on Federal Bullets F001 and F002

| | log(concentration) on F001 | | | | | |
|---|---|---|---|---|---|---|
| | ICP-Sb | ICP-Cu | ICP-Ag | ICP-Bi | ICP-As | ICP-Sn |
| a | 10.28452 | 5.65249 | 4.15888 | 2.77259 | 7.25488 | 7.51861 |
| b | 10.29235 | 5.61677 | 4.30407 | 2.77259 | 7.29980 | 7.51643 |
| c | 10.27505 | 5.64545 | 4.18965 | 2.77259 | 7.24708 | 7.48997 |
| mean | 10.28397 | 5.63824 | 4.21753 | 2.77259 | 7.26725 | 7.50834 |
| SD | 0.00866 | 0.01892 | 0.07650 | 0.00000 | 0.02845 | 0.01594 |
| | log(concentration) on F002 | | | | | |
| | ICP-Sb | ICP-Cu | ICP-Ag | ICP-Bi | ICP-As | ICP-Sn |
| a | 10.27491 | 5.62762 | 4.33073 | 2.77259 | 7.29506 | 7.52994 |
| b | 10.26928 | 5.63121 | 4.20469 | 2.77259 | 7.27170 | 7.49387 |
| c | 10.27135 | 5.64191 | 4.34381 | 2.70805 | 7.28001 | 7.47760 |
| mean | 10.27185 | 5.63358 | 4.29308 | 2.75108 | 7.28226 | 7.50047 |
| SD | 0.00285 | 0.00743 | 0.07682 | 0.03726 | 0.01184 | 0.02679 |
| $s_{j,pool}$ | 0.0192 | 0.0200 | 0.0825 | 0.0300 | 0.0432 | 0.0326 |
| RMD $s_{j,pool}$ | 0.631 | 0.233 | –0.916 | 0.717 | –0.347 | 0.241 |

distribution), and that a "CIVL" has been defined to be as small a volume as is needed to ensure that the variability of the elemental concentrations within this volume is much smaller than the measurement uncertainty (i.e., within-lot variability is much smaller than σ). It does not depend on any assumptions about the distribution of elemental concentrations in the general population of bullets, for which we have no valid data sets that would allow statistical inference. Probabilities such as the FBI's claim of "1 in 2,500" are inappropriate when based on a data set such as the 1,837-bullet data set; as noted in Section 3.2, it is not a random collection of bullets from the population of all bullets, or even from the complete 71,000+ bullet data set from which it was extracted.

The use of either $0.63s_{j,pool}$ or $1.07s_{j,pool}$ (requiring $\bar{x}_j$ and $\bar{y}_j$ to be within 1.0 to 1.5 times the measurement uncertainty), might seem too demanding when only three pairs of bullets among 854 bullets (subset of the 1,837-bullet data set in which all seven elements were measured) showed differences of less than or equal to 1 SD on all seven elements (eight pairs of bullets had maximal *RMD*s of 1.5). However, as noted in the paragraph describing the data set, the 1,837 bullets were selected to be unrelated (Ref. 6), and hence do not represent, in any way, any sort of random sample from the population of bullets. We cannot say on the basis of this data set, how frequently two bullets manufactured from different sources may have concentrations within 1.0. We do know that such instances can occur. A carefully designed study representative of all bullets that might exist now or in the future may help to assess the distribution of differences between mean concentrations of different bullets and may lead to a different choice of the constant, depending on the level of $\delta/\sigma$ that the procedure is designed to protect. Constants for other values of the per-element FPP (0.01, 0.05, 0.10, 0.20, 0.222 and 0.30) and $\delta$ (0.25, 0.50, 1.0, 1.5, 2.0, and 3.0), for $n = 3$ and $n = 5$, are given in Appendix F. See also Box K.1

## 4.3 Hotelling's $T^2$

A statistical test procedure that is designed for comparing two sets of 7 sample means simultaneously rather than 7 individual tests, one at a time, as in the previous section, uses the estimated covariance matrix for the measurement errors. The test statistic can be written

$$T^2 = n\bar{d}'S^{-1}\bar{d} = n(\bar{d}/s)'R^{-1}(\bar{d}/s)$$

where:

- $n$ = number of measurements in each sample mean (here, $n = 3$).
- $p$ = number of elements being measured (here, $p = 7$).
- $\bar{d} = \bar{x} - \bar{y}$ = mean difference in the seven elements expressed as a column vector of length $p$ ($\bar{d}'$ = row vector of length $p$).
- **s** = vector of SDs in measuring the elements (length $p$).
- $S^{-1}$ = inverse of the estimated matrix of variances and covariances among the measurement errors (seven rows and seven columns).
- $R^{-1}$ = inverse of the estimated matrix of correlations among the measurement errors (seven rows and seven columns).
- $\nu$ = number of degrees of freedom in estimating $S$, the matrix of variances and covariances (here, 2 times the number of bullets if three measurements are made of each bullet).

Under the assumptions that

- the measurements are normally distributed (for example, if lognormal, then the logarithms of the measurements are normally distributed),
- the matrix of variances and covariances is estimated very well, using $\nu$ degrees of freedom (for example, $\nu = 200$, if three measurements are made on each of 100 bullets and the variances and covariances within each set of three measurements are pooled across the 100 bullets), and
- the bullet means truly differ by $\delta/\sigma = 1$ in each element,

$[\nu + 1 - p)/(p\nu)]T^2$ should not exceed a critical value determined by the noncentral F distribution with $p$ and $\nu$ degrees of freedom and noncentrality parameter given by $n(\delta/\sigma)R^{-1}(\delta/\sigma) = 3(\delta/\sigma)$ times the sum of the elements in the inverse of the estimated correlation matrix (Ref. 16, pp. 541–542). When $p = 7$ and $\nu = 400$ degrees of freedom, and using the correlation matrix estimated from

## BOX K.1
## True Matches and Assessed Matches

The recommended statistical test procedure for assessing a match will involve the calculation of the sample means from the measurements (transformed via logarithms) on the CS and PS bullets and a pooled standard deviation (as an estimate of the measurement uncertainty). If the sample means on all seven elements are "too close," relative to the variability that is expected for a difference between two sample means, then a "match" is declared. "Too close" is determined by a constant that arises from either a non-central *t* distribution, if a *t*-test on each individual element is performed, or a non-central F distribution, if Hotelling's $T^2$ test is used, where the relative mean differences are combined and weighted in accordance with the correlation among the seven measurement errors.

Two types of questions may be posed. The first type involves conditioning on the difference between the bullet means: Given that two bullets really did come from the same CIVL (compositionally indistinguishable volume of lead), what is the probability that the statistical test procedure correctly claims "match"? Similarly, given two bullets that are known to have come from different CIVLs, what is the probability that the test correctly claims "no match"? Stated formally, if δ represents the vector of true mean differences in the seven elemental concentrations, and if "*P(A|B)*" indicates the probability of A, given that B holds, then these first types of questions can be written: What are *P(claim "match"* | δ = 0) and *P(claim "nonmatch"* | δ = 0) (where these two expressions sum to 1 and the second expression is the false non-match probability), and what are *P(claim "match"* | δ > 0) and *P(claim "nonmatch"* | δ > 0) (again where these two expressions sum to 1, and the first expression is the false match probability )?

In other words, one can ask about the performance of the test, given the true connection between the bullets. Using a combination of statistical theory and simulation, these probabilities can be estimated for the FBI's current match procedures as well as for the alternative procedures recommended here.

The second type of question that can be asked reverses terms and now involves conditioning on the assessment and asking about the state of the bullets. One of the two versions of this type of question is: Given that the statistical test indicates "match", what is the probability that the two bullets came from the same CIVL?

The answer to these questions depends on several factors. First, as indicated in Chapter 3, we cannot *guarantee* uniqueness in the mean concentrations of all seven elements simultaneously. Uniqueness seems *plausible*, given the characteristics of the manufacturing process and the possible changes in the industry over time (e.g., very slight increase in silver concentrations over time). But uniqueness cannot be assured. Therefore, at best, we can address only the following modified question: "If CABL analysis indicates "match," what is the probability that these two bullets were manufactured from CIVL's that have the same mean concentrations on all seven elements, compared with the probability that these two bullets were manufactured from CIVLs that differ in mean concentration on one or more of the seven elements?"

Using the notation above, this probability can be written: *P*(δ = 0 | *claim*

*continued*

**BOX K.1 continued**

*"match")*, which is 1 – *P(*$\delta$ > 0 | *claim "match")*. Similarly, one can ask about the *P(*$\delta$ = 0 | *claim "nonmatch")*, which is 1 – *P(*$\delta$ > 0 | *claim "nonmatch")*.

By applying Bayes' rule (Ref. 8),

*P(*$\delta$ = 0 | *claim "match")* = *P(claim "match"* | $\delta$ = 0*)P(*$\delta$ = 0*)* / *P(claim "match")*

and

*P(*$\delta$ > 0 | *claim "match")* = *P(claim "match"* | $\delta$ > 0*)P(*$\delta$ > 0*)* / *P(claim "match"*

The ratio between these two probabilities, i.e. *P(*$\delta$ = 0 | *claim "match")*/ *P(*$\delta$ > 0*)* | *claim "match")* is equal to: *P(claim "match"* | $\delta$ = 0*)P(*$\delta$ = 0*)* / *P(claim "match"* | $\delta$ > 0*)P(*$\delta$ > 0*)* (*)

One might reflect, "Given that the CABL analysis indicates "match," what is the probability that the bullets came from populations with the same mean concentrations, compared to the probability that the bullets came from different populations?" A large ratio might be strong evidence that the bullets came from CIVLs with the same mean concentrations. (In practice, one might allow a small $\delta_0$ so that "$\delta < \delta_0$" is effectively a "match" and "$\delta > \delta_0$" is effectively a "non-match"; the choice of $\delta_0$ will be discussed later, but for now we take $\delta_0 = 0$.) The above equation shows that this ratio is actually a product of two ratios, one *P(claim "match"* | $\delta$ = 0*)* / *P(claim "match"* \ *(*$\delta$ > 0*)*, which can be estimated as indicated above through simulation, and where a larger ratio indicates a more sensitive test, and a second ratio *P(*$\delta$ = 0*)* |*P(*$\delta$ > 0*)* which depends on the values of the mean concentrations across the entire universe of CIVLs (past, present, and future). Section 3 below estimates probabilities of the form of the first ratio and shows that this ratio exceeds 1 for all tests, but especially so for the alternative procedures recommended here. However, the second ratio is unknown, and, in fact, depends on many factors:

1. the consistency of elemental concentration within a CIVL ("within-CIVL homogeneity");
2. the number of bullets that can be manufactured from such a homogeneous CIVL;
3. the number of CIVLs that are analytically indistinguishable from a given CIVL (in particular, the CIVL from which the CS bullet was manufactured);
4. the number of CIVLs that are *not* analytically indistinguishable from a given CIVL.

These factors will vary by type of bullet, by manufacturer, and perhaps by locale (i.e., more CIVLs are readily accessible to residents of a large metropolitan area than to those in a small urban town).

This appendix analyzes data made available to the Committee in an attempt to estimate a frequency distribution for values of $\delta$ in the population, which is needed for the probabilities in the second ratio above. However, as will be seen, these data sets are biased, precluding unbiased inferences. In the end, one can conclude only that *P(*$\delta$ > 0 | *claim "match")* > *P(*$\delta$ = 0), i.e., given the results of a test that suggests "match," the probability that the two bullets came from the same CIVL is higher than this probability if the two bullets had not been measured at all. This, of course, is a weak statement. A stronger statement, namely, that the ratio of the probabilities in (*) exceeds 1, is possible only through a carefully designed sampling scheme, from which estimates, and corresponding confidence intervals, for the probability in question (*), can be obtained. No such unbiased information is currently available. Consequently, the recommended alternative statistical procedures (Hotelling's $T^2$ test and successive individual Student's *t* tests on the seven elements separately) consider only the measurable component of variability in the problem, namely, the measurement error, and not the other sources of variability (within-CIVL and between-CIVL variability), which would be needed to estimate this probability.

We note as a further complication to the above that the linkage between a "match" between the CS and PS bullets and the inference that these two bullets came from the same CIVL depends on how a CIVL is defined. If a CS bullet is on the boundary of a CIVL, then the likelihood of a match to bullets outside a CIVL may be much higher than if a CS bullet is in the middle of a CIVL.

the Federal data (which measured six of the seven elements with ICP-OES; see Appendix F) and assuming that the measurement error on Cd is 5% and is uncorrelated with the others, this test procedure claims analytically indistinguishable (match) only if $T^2$ is less than 1.9 ($\delta/\sigma = 1$ for each element) and claims analytically indistinguishable (weak match) only if $T^2$ is less than 6.0 ($\delta/\sigma = 1.5$ for each element), to ensure an overall FPP of no more than 0.0004 (1 in 2,500).[1] (When applied to the log(concentrations) on Federal bullets F001 and F002 in Table K.14, the value of Hotelling's $T^2$ statistic, using only six elements, is 2.354, which is small enough to claim "analytically indistinguishable" when $\delta/\sigma = 1.0$ and the overall FPP is 0.002, or 1 in 500.)

The limit 1.9 depends on quite a large number of assumptions. It is indeed more sensitive if the correlation among the measurement errors is substantial (as it may be here for at least some pairs of elements) and if the differences in element concentrations tend to be spread out across all seven elements rather than concentrated in only one or two elements. However, the validity of Hotelling's $T^2$ test in the face of departures from those assumptions is not well understood. For example, the limit 1.9 was based on an estimated covariance matrix from one set of 200 bullets (Federal) from one study conducted in 1991, and the inferences from it may no longer apply to the current measurement procedure. Also, although Hotelling's $T^2$ test is more sensitive at detecting *small* differ-

[1]For an overall FPP of 0.002 (1 in 500), the test would claim "match" or "weak match" if $t^2$ does not exceed 1.9 or 8.1, respectively. For an overall FPP of 0.01 (1 in 100), the test would claim "match" or "weak match" if $t^2$ does not exceed 4.5 or 11.5, respectively.

ences in concentrations in *all* elements, it is less sensitive than the individual $t$ tests if the main cause of the difference between two bullets arises from only one fairly large difference in one element. (That can be seen from the fact that, if the measurement errors were independent, $T^2/p$ reduces to the average of the squared two-sample $t$ statistics on the $p = 7$ separate elements, so one large difference is spread out across the seven dimensions, causing $[\nu + 1 - 7)/\nu]T^2/p$ to be small and thus to declare a match when the bullets differ quite significantly in one element.) Many more studies would be needed to assess the reliability of Hotelling's $T^2$ (for example, types of differences typically seen between bullet concentrations, precision of estimates of the variances and covariances between measurement errors, and departures from (log)normality).

### 4.4 Use of T Tests in Court

One reason for the authors' recommendation of seven individual equivalence $t$ tests versus its multivariate analog based on Hotelling's $T^2$, is the familiarity of the form. Student's $t$ tests are in common use and familiar to many users of statistics; the only difference here is the multiplier ("0.63" for "match" or "1.07" for "weak match," instead of "2.0" in a conventional $t$ test, $\alpha = 0.05$). The choice of FPP, and therefore the determination of $\delta$, could appear arbitrary to a jury and could subject the examiner to a difficult cross examination. However, the choice of $\delta$ is in reality no more arbitrary than the choice of $\alpha$ in the conventional $t$ test—the "convention" referred to in the name is in fact the choice $\alpha = 0.05$, leading to a "2.0-sigma" confidence interval. The conventional $t$ test has the serious disadvantage that it begins from the null hypothesis that the crime scene bullet and the suspect's bullet match, that is, it starts from the assumption that the defendant is guilty ("bullet match") and sets the probability of falsely assuming that the guilty person is innocent to be .05. This drawback could be overcome by computing the complement of the conventional $t$ test Type II error rate (the rate at which the test fails to reject the null hypothesis when it is false, which in this case would be the false positive result) for a range of alternatives to the null hypothesis and expressing the results in a power curve in order to judge the power of the test. However, this is not as appealing from the statistician's viewpoint as the equivalence $t$ test. (It is important to note that the standard $t$ test-based matching error rate will fluctuate by bullet manufacturer and bullet type. This is due to the fact that difference among CABLs are characteristic of manufacturer and bullet type.)

Table K.15 presents a comparison of false positive and false negative rates using the FBI's statistical methods, and using the equivalence and conventional $t$-tests.

It is important to note that this appendix has considered tests of a "match" between a single CS bullet and a single PS bullet. If the CS bullet were com-

TABLE K.15 Simulated False-Positive and False-Negative Probabilities Obtained with Various Statistical Testing Procedures

| | Composition Identical $\delta = 0$ | Composition Not Identical $\delta = 1.5$ |
|---|---|---|
| CABL claims "match" | | |
| | True Positive | False Positive |
| FBI-2SD | 0.933 | 0.571 |
| FBI-rg | 0.507 | 0.050 |
| Conv t | 0.746 | 0.065 |
| Equiv-t (1.3) | 0.272 | 0.004 |
| HotelT$^2$ (6.0) | 0.115 | 0.001 |
| CABL claims "no match" | | |
| | False Negative | True Negative |
| FBI-2SD | 0.067 | 0.429 |
| FBI-rg | 0.493 | 0.948 |
| Conv t | 0.254 | 0.935 |
| Equiv-t (1.3) | 0.728 | 0.996 |
| HotelT$^2$ (6.0) | 0.885 | 0.999 |

Note: Simulated false-positive and false-negative probabilities obtained with various statistical testing procedures. Simulation is based on 100,000 trials. In each trial, 3 measurements on seven elements were simulated from a normal distribution with mean vector $\mu_x$, standard deviation vector $\sigma_x$, and within-measurement correlation matrix R, where $\mu_x$ is the vector of 7 mean concentrations from one of the bullets in the 854-bullet data set, $\sigma_x$ is the vector of 7 standard deviations on this same bullet, and R is the within-measurement correlation matrix based on data from 200 Federal bullets (see Appendix F). Three further measurements on seven elements were simulated from a normal distribution with mean vector $\mu_y = \mu_x + k\sigma_x$, with the same standard deviation vector $\sigma_x$, and the same within-measurement correlation matrix R, where $\mu_y$ is the same vector of mean concentrations plus an offset equal to $k$ times the measurement uncertainty in each element. The simulated probabilities of each test (FBI 2-SD overlap, FBI range overlap, conventional $t$, equivalence $t$) equal the proportions of the 100,000 trials in which the test claimed "match" or "no match" (i.e., the sample means on all 7 elements were within 0.63 of the pooled estimated of the measurement uncertainty in measuring that element). For the first column, the simulation was run with $k = 0$ (i.e., mean concentrations are the same); for the second column, the simulation was run with $k = 1$ (i.e., mean concentrations differ by 1.5 times the measurement uncertainty). With 100,000 trials, the uncertainties in these simulated probabilities (two standard errors) do not exceed 0.003. Note that $\sigma_x$ is the measurement error, and we can consider this to be equal to $\sqrt{\sigma_l^2 + \sigma_{inh}^2}$, where $\sigma_l$ is the measurement uncertainty and $\sigma_{inh}$ is uncerainty due to homogeneity.

pared with, say, 5 PS bullets, all of which came from a CIVL whose mean concentrations differed by at least 1.5 times the measurement uncertainty ($\delta = 1.5\sigma$), then, using Bonferroni's inequality, the chance that the CS bullet would match *at least one* of the CS bullets could be as high as five times the nominal FPP (e.g., 0.01, or 1 in 100, if the "1 in 500" rate were chosen). Multiplying the current false positive rates for the FBI 2-SD-overlap and range-overlap procedures shown in Table K.15 by the number of bullets being tested results in a very high probability that at least one of the bullets will appear to "match," simply by chance alone, even when the mean CIVL concentrations of the two bullets differ by 1.5 times the measurement uncertainty 3–7%). The small FPP for the equivalence $t$ test results in a small probability that some CS bullet will match the PS bullet by chance alone, so long as the number of PS bullets is not very large.

## REFERENCES


1. Laboratory Chemistry Unit. Issue date: October 11, 2002. *Unpublished* (2002).
2. Peele, E. R.; Havekost, D. G.; Peters, C. A.; Riley, J. P.; Halberstam, R. C.; and Koons, R. D. USDOJ (ISBN 0-932115-12-8), 1991, 57.
3. Peters, C. A. *Foren. Sci. Comm.* 2002, *4(3).* <http://www.fbi.gov/hq/lab/fsc/backissu/july2002/peters.htm> as of Aug. 8, 2003.
4. 800-bullet data set provided by FBI in email from Robert D. Koons to Jennifer J. Jackiw, February 24, 2003.
5. 1,837-bullet data set provided by the FBI. (CD) Received by committee May 12, 2003.
6. Koons, R. D. Personal communication to committee. (CD) Received by committee May 12, 2003. Description of 1,837-bullet data set.
7. Randich, E.; Duerfeldt, W.; McLendon, W.; and Tobin, W. *Foren. Sci. Int.* 2002,127, 174–191.
8. Carriquiry, A.; Daniels, M.; and Stern, H. "Statistical Treatment of Case Evidence: Analysis of Bullet Lead," *Unpublished report,* Dept. of Statistics, Iowa State University, 2002.
9. Grant, D. M. Personal communication to committee. April 14, 2003.
10. Koons, R. D. Personal communication to committee via email to Jennifer J. Jackiw. March 3, 2003.
11. Koons, R. D. "Bullet Lead Elemental Composition Comparison: Analytical Technique and Statistics." Presentation to committee. February 3, 2003.
12. Vardeman, S. B. and Jobe, J. M. *Statistical Quality Assurance Methods for Engineers*; Wiley: New York, NY 1999.
13. Wellek, S. *Testing Statistical Hypotheses of Equivalence;* Chapman and Hall: New York, NY 2003.
14. Owen, D.B. "Noncentral t distribution" in *Encyclopedia of Statistical Sciences, Volume 6*; Kotz, S.; Johnson, N. L.; and Read, C. B.; Eds.; Wiley: New York, NY 1985, pp 286–290.
15. Tiku, M. "Noncentral F distribution" in *Encyclopedia of Statistical Sciences, Volume 6*; Kotz, S.; Johnson, N. L.; and Read, C. B.; Eds.; Wiley: New York, NY 1985, pp 280–284.
16. Rao, C.R., *Linear Statistical Inference and Its Applications*; Wiley, New York, NY 1973.